Utopian Discourses Across Cultures

Miriam Bait / Marina Brambilla / Valentina Crestani (eds.)

Utopian Discourses Across Cultures

Scenarios in Effective Communication to Citizens and Corporations

PL ACADEMIC RESEARCH

Bibliographic Information published by the Deutsche Nationalbibliothek
The Deutsche Nationalbibliothek lists this publication in the Deutsche Nationalbibliografie; detailed bibliographic data is available in the internet at http://dnb.d-nb.de.

Cover image: © ibreakstock - Fotolia

Library of Congress Cataloging-in-Publication Data
Names: Bait, Miriam, 1959- editor. | Brambilla, Marina Marzia, 1973- editor. | Crestani, Valentina, 1983- editor.
Title: Utopian discourses across cultures : scenarios in effective communication to citizens and corporations / Miriam Bait, Marina Brambilla, Valentina Crestani (eds.).
Description: Frankfurt am Main ; New York : Peter Lang, [2016]
Identifiers: LCCN 2016016343 | ISBN 9783631666838
Subjects: LCSH: Utopias. | Communication--Social aspects.
Classification: LCC HX806 .U79324 2016 | DDC 335/.02--dc23 LC record available at https://lccn.loc.gov/2016016343

ISBN 978-3-631-66683-8 (Print)
E-ISBN 978-3-653-06174-1 (E-Book)
DOI 10.3726/978-3-653-06174-1

PL Academic Research is an Imprint of Peter Lang GmbH.

Peter Lang – Frankfurt am Main · Bern · Bruxelles · New York · Oxford · Warszawa · Wien

This publication has been peer reviewed.

www.peterlang.com

Contents

3. Education and Communication

Foreword

This volume consists of ten papers, most of which were originally presented at the workshop "Scenari utopistici nella comunicazione pubblica e d'impresa" ("Utopian scenarios in public and corporate communication") in December 2014. The workshop was held within the framework of an interdisciplinary and inter-departmental project on the concept of utopia, funded by *Università degli Studi di Milano* (Milan).

The term *utopia*, as coined by Thomas More in 1516, contains an inherent semantic ambiguity: 'utopia' can be defined as *eu topos* (good place) or *ou topos* (no place). It is an incredibly powerful, though polysemic notion for imagining and producing a "better world" that has always been associated with a place, a territory, a circumscribed space where utopian ideals are realized. If the bounds of a utopia were to be overcome, that world could become universal. Such a fascinating notion can, of course, be related to the time and place where it is used. In the light of such (non) definitions, a number of intersections, representations, re-elaborations, and re-contextualizations allows for analyses using different critical methods, and a variety of tools from different disciplines. The rich and diverse set of critical approaches, ranging from discourse analysis and intercultural studies to the social sciences, perfectly mirrors the wide variety of ways in which the term *utopia* has been defined and used regarding content, form, and function.

The authors of the following essays are experts in a number of disciplines, including sociology, economics, computer science, literature and linguistics. The papers focus on the idea of utopia by considering it from the author's specific disciplinary perspective. Accordingly, each essay offers ideas unique to their disciplinary point of view, and taken together the articles comprise an interdisciplinary volume. Structured around three thematic sections, the articles in this collection have been organized as follows:

1. Society and Communication.
2. Economics and Communication.
3. Education and Communication.

Papers by Miriam Bait, Federico Boni and Paola Bozzi are included in the first section "Society and Communication".

Bait presents observations and findings gleaned from an exploratory analysis conducted on a corpus of texts taken from public websites of city councils in the United Kingdom. She investigates the discursive practices of the public discourse deployed by local authorities in their communication to the public, analyzing whether, and to what extent, a new managerial logic in the public sector has been adopted in order to reach the utopian goal of effective integration between citizens and governments.

In his essay, Boni aims to investigate a particular kind of utopia of communication: the idea of communication as a positive value. Following Mannheim's work, the concept of utopia is turned into a situational and socially conditional concept. This leads to a semiotic concept of 'myth' in which 'myth' refers to an unarticulated chain of associated concepts and discourses by which members of a society understand certain topics.

In her paper, Bozzi explores the politics of place branding and the process of urban reinvention in Berlin after 2001. In the context of the dramatic socio-economic restructuring processes and physical transformation of the city following the fall of the Wall, the new Mayor of Berlin, Klaus Wowereit, tried to replace Berlin's financial squalor with the beauty and joy of sex, sketching the utopian scenario of a creative city, which "is poor but sexy".

The second section, "Economics and Communication", includes essays by Paola De Vecchi Galbiati, Raffaella Folgieri, Angela Lupone and Gloria Regonini.

De Vecchi Galbiati's article provides an interesting overview of the main outcomes obtained in managing change management projects with an innovative approach that imitates natural organisms' structure and behaviors. The apparently utopian result of this study is that the participants started seeing themselves as "natural and cultural environments" able to adapt and change in several situations. After a brief introduction about what adaptive systems are and how they work, the article provides a case study detailing the evolution of three real communities that have applied these approaches based on self-adaptation processes and achieved successful results.

Folgieri highlights the impact of advanced emerging technologies on the current economic situation, with a particular focus on artificial intelligence. Recalling Keynes' famous speech on future economy, she offers an overview of potential technological factors that may be contributing to accelerating the present economic crisis. Folgieri also analyzes the question of whether recent progress might represent an opportunity or an obstacle in the search for a new system of economic balance.

Lupone's article deals with the topic of fair and free trade by comparing it with the main theme explored in French director Serreau's film *The Green Beautiful* (*La Belle Verte*). The Green Beautiful is a planet very much like planet Earth, but a utopian society has emerged there. Lupone's perspective on fair trade is twofold, as she manages to counter the harshness of daily reality with the hope of a better future.

Gloria Regonini's study demonstrates the necessity of adopting multiple disciplinary perspectives when complex economic problems with a long history of failures come into play, such as the measures that have been taken in Italy to simplify the bureaucratic requirements for small and medium enterprises. In such cases, complexity – when it is not recognized and analyzed through a variety of paradigms – will take its revenge on solutions that seem easy upon first sight but are, in truth, completely ineffective. In relation to regulative policies, the difficulties of learning from the best foreign experiences are aggravated by misalignments of meaning between English and Italian terms that seem to be very similar but are actually very different, like "policy" and "politica", or "regulation" and "regolamentazione". Such mistranslations are not merely linguistic mistakes, but stem from different visions of the public sphere and its institutions.

In the third section entitled "Education and Communication", articles by Marina Brambilla / Valentina Crestani and by Giampietro Gobo investigate the concept of utopia by referring to the two main groups of actors in academic settings: (future) lecturers and students. The paper by Claudia Gualtieri is also included in this section, as it focuses on Indian Residential Schools in Canada.

Brambilla and Crestani research the process of communication between the university and its students by analyzing the online presentation of the German *Universität Duisburg-Essen*. They concentrate on the relationship between the graphic and verbal components of the website, which is not always a one-to-one connection. In some cases, the words underscore utopian concepts like meritocracy and gender equality, while photographs emphasize other aspects.

Gobo examines disparities in the university setting, suggesting a thought-provoking relation between care activities and scientific production in academic contexts, and to what extent the former might slow down the latter. The author suggests that one affirmative action, in academic environments, might be the use of the Care Factor, a tool used to weigh the scientific productivity of a candidate bearing the child-rearing.

Gualtieri's contribution takes up the findings presented in the *Truth and Reconciliation Commission (TRC) Findings on Indian Residential Schools in Canada*, held in June 2015, to explore the issue of indigeneity in contemporary Canada. This essay examines the colonial history of residential schools for indigenous people, how national politics dealt with the issue, and describes the social and cultural consequences of the unequal educational system in Canada. The debate triggered by the release of the TRC's findings speaks to the need for urgent political action to establish a new, just, and more stable foundation for a more inclusive – and maybe less utopian – society.

The variety of contributions contained in this volume reflects the different perspectives from which the phenomenon of utopia is 'realized' in communication situations. The oxymoron of the 'realization of utopia' is indeed present in communication as a verbal and non-verbal process involving many aspects of people's lives as part of a linguistically and culturally defined community. Society, in the following papers, consists of a system in which internal and external relations and regulations are analyzed and determined according to cultural, economic and educational parameters. The complexity of society leads to the emergence of dystopian situations in a system that purports to be based on utopian ideals, but where these ideals often remain abstract ideals – or become their opposite.

Miriam Bait, Marina Brambilla, Valentina Crestani

1. Society and Communication

Miriam Bait

The United Kingdom Is(a)land of Utopia: Self-Representation of City Councils and Communicative Strategies towards Citizens

1. Introduction

The concept of utopia has always been associated with a place, a territory, a confined space in which the utopia is realized. And once its boundaries have been overstepped, it could become universal. Utopia inevitably references a social ideal and offers a model of coexistence, imagining how men should or could organize their lives.

A utopian model of society and the institutions upon which this society should stand becomes visible in the communicative acts through which public administrations speak about themselves. While the great literary utopias of the past seem to have faded into the background, they reemerge in today's institutional communication.

The utopia that I will try to outline in these pages is, in fact, a contemporary utopia, one concerning "public communication," an expression that may sound like an oxymoron, especially for historical reasons, since institutions and governments have generally not been known for excelling in their communication with the public.

But the profound transformation that has taken place in recent decades in western public administrations has greatly influenced the ways in which public bodies can communicate with and thus relate to the citizens they serve. At the center of what has been called "new public management", "entrepreneurial Government", or "neo-managerialism" (Hackney and McBride 1995; Terry 1998) is the adoption of good management and, above all, effective marketing strategies focused on the needs of the community. This increasingly business-like orientation, both in terms of practices and of discursive strategies, addresses citizens who are no longer considered to be simple users, but rather stakeholders and customers to keep satisfied.

2. Background and Aims

The utopia of the contemporary city becomes a common city project, an opportunity and a means of social change to be built upon the basis of joint decisions.

In this dynamic of change, new technologies have played a key role and have radically changed the modes of interaction between public agencies and companies, as well as the ways in which institutions can relate to citizens, as they not only help to implement the self-representation of institutions for propaganda purposes, but encourage participation, the creation of networks and, ultimately, greater democratization and better governance (Wilhelm 2000; Coleman 2001; West 2004; Karakaya 2005).

The institution or, in this analysis, the municipality, becomes visible and is forced to submit, consciously and strategically, its realistic (or utopian) vision of the city, a socio-political transposition of the utopian ideal, of imaginary and literary places. The result is a continuous (electronic) democracy based on the discussion, the recruitment and the aggregation of collective energies through the web.

The Internet has potentially disintermediated most of the informational actions between institutions and citizens. Informing is not enough: the actions of institutions should be supported by marketing and communication strategies to guide the customer / user towards their own structure by sharing their own point of view with them, and at the same time impacting the guidelines and perception of what institutions are and what they are doing. Online presence "help[s] build an image, foster an identity, drive social and economic development, and fortify the cohesion of a local community" (Jeffres and Lin 2006: 957). The unique nature of online discourse transforms Habermas' notion of the public sphere (Habermas: 2006) into a revived version, a sphere where public opinion and consensus are formed through communicative action, thus paving the road for a democratic utopia, "a discursive arena that is home to citizen debate, deliberation, agreement and action" (Villa 1992: 712).

3. Data and Methodology

In order to investigate how strategic objectives deployed by institutions are discursively represented, I have selected the websites of four cities or district councils in the United Kingdom (UK) for the purposes of comparison (Chichester, Exeter, Leicester, Norwich)[1]. Texts describing the institutions' present commitment (e.g. mission statement) and future plans (e.g. vision statement) were taken from the respective websites and stored on a computer to provide text file formats for the current analysis, yielding an entire dataset of approximately 150,000 words.

---

1 <http://chichestercity.gov.uk/>, <http://exeter.gov.uk/>, <http://www.leicester.gov.uk/>, <http://www.greaternorwichgrowth.org.uk/>. All websites were accessed from 13/11/2014 to 16/10/2015.

Sections labelled as "mission statements", as well as other similar sections under different headings such as such as "vision", "aims", "priorities", "business responsibilities", "values", and so forth have been taken into consideration. These texts were examined to identify possible common traits which may provide evidence of an ideal collective identity to be constructed with the help of different subjects (public authorities and citizens).

The number of words per individual dataset is shown in the following table:

Table 1: Number of words per dataset

City or District Council	Number of Words
Chichester	32,880
Exeter	33,417
Leicester	25,821
Norwich	53,062

The methodological approach I use draws on principles and analytical tools from the tradition of discourse analysis. Particular attention is given to linguistic features such as evaluative language (Hunston and Thompson 2001) and specific lexical choices revealing a contamination between two different domains (private and public), thus outlining a dual vision of utopia, based on the principles of justice and equality, but promoted as making use of market-oriented strategies.

4. The Place of Utopia: Discursive Strategies And Linguistic Features

The British websites examined in this study display a range of different textual genres. In the context of the kind of strategic communication that was a prerogative of businesses and private organizations in the past, these genres refer to the mission of institutions (Bait 2008, 2009). Terms like vision, values, aims, priorities or business responsibilities are very common, as we can see in the following examples:

> *Extending horizons for the whole community.* Our vision is quite simply: to improve the lives of those in our community, and to maintain and enhance the environment. Our values are what govern us as an organisation and we as people serving the community and by values we mean being responsive, open, fair, supportive, caring. (Chichester District Council)

> *Our city, Our Future*. We want to maintain and build upon Exeter's position as a regional capital and work with all sectors of the local community to provide a healthy, prosperous and safe place for people to live, work and visit. (Exeter City Council)
>
> *Making Leicester more attractive for our diverse communities to live, work and invest in*. Our values are: building trust, valuing staff, cultivating leadership and delivering quality. (Leicester City Council)
>
> *Putting the city and its people first*. Everything we ever do as an organisation, whether in teams or as individuals, will be done with our core values in mind: Pride, Accountability, Collaboration, Excellence. (Norwich City Council)

These extracts clarify the intentions of the institutions to depict themselves as actors for change and as efficient service-providers, capable of responding to the needs of citizens and ensuring their satisfaction:

> […] a leading authority, a trusted partner and an organisation that is both outward looking and good to work with […] (Norwich City Council)

The reference to partnership and cooperation highlights that the representation of 'ideal' institutions also entails the representation of 'ideal' citizens. To actually be realized, the action taken by the institution must be accompanied and supported by citizens, who, in turn, become not only users of services but also actors, co-builders and propagators of change.

4.1 Actors And Actions

At the linguistic level, textual analysis allows us to acknowledge that the partnership between institutions and citizens is confirmed and emphasized by the use of strategic pronominal references (Benveniste 1966; De Fina 1995; Wilson 1990). On the one hand, the exclusive *we* – and the attribute adjective 'our' – present the institution as committed to the development of positive actions and the implementation of projects:

> At Chichester District Council, <u>we</u> try to make sure that you always receive a first class service. We hope you will never need to complain. (Chichester District Council, emphasis added)
>
> <u>We</u> remained focused on <u>our</u> two main challenges… (Leicester City Council, emphasis added)
>
> Exeter City Council intends to meet these challenges in the next 15 years. It will not be easy, the future will demand that <u>we</u> make things happen in the right way and <u>we</u> may have to adjust our plans in response to changing times and world events.
>
> <u>We</u> will strive to do things well and look for ways to innovate and improve… (Norwich City Council, emphasis added)

On the other hand, the use of the inclusive *we* helps to construct a shared social identity in which both producers (institutions) and users (citizens) play a role, as seen in the following example:

> The Core Strategy represents the Council's vision and objectives to take us up to 2026 and [...] by working together we can achieve them. (Exeter City Council, emphasis added)

Moreover, the pronoun *you* introduces a significant interpersonal component in the text (Halliday 1994: 69–158), a sign of a wider process of "informalization" (Wouters 1986; Featherstone 1991) and "conversationalization" (Fairclough 1995) of public discourse. In particular, directly addressing the reader with 'you' simply does not build a relationship of "equality, solidarity [...] intimacy" (Fairclough 2001: 52), but also identifies readers as the ultimate addressees of the institution's initiatives, as is apparent in the following example:

> We hope you will never need to complain. (Chichester District Council, emphasis added)

More importantly, the fact that the pronoun *you* is the main deictic marker of dialogue strengthens the impression that the text producer is engaging the reader in a real dialogic conversation by means of questioning:

> Would you like us to produce an Annual Report, available on our website, which would give a summary on our performance and our financial position? What would you like to see featured in future editions of an Annual Report? (Chichester District Council, emphasis added)

Defining issues of public concern in the public interest helps to create a real participatory culture for which the community becomes an actor of change in civic affairs:

> [...] other services are not statutory, but the community, through its elected representatives, deem them to be highly desirable and want them carried out. (Chichester District Council, emphasis added)

> The Core Strategy represents the Council's vision and objectives to take us up to 2026 and I hope that, by working together we can achieve them. (Exeter City Council, emphasis added)

> The city council has a civic leadership role and our elected councillors have a mandate and responsibility to represent and work on behalf of their communities for the broader interests of the city. (Norwich City Council, emphasis added)

Citizens are not seen as isolated entities, but as members of an interdependent community where transparency in public administration and urban democracy becomes possible.

> We are keen for as many people as possible to get involved in the process and will be in contact with key groups including Parish and Town Councils.
>
> We use a variety of methods to consult and involve the local community, such as surveys, focus groups, exhibitions or consultation leaflets. (Chichester District Council, emphasis added)

In addition to the aforementioned lexical item 'together', the emphasis on the concept of community and partnership is further strengthened by the frequent use of the terms *everyone*, *each of us*, and *all*, which express the idea of sharing resources, information, experiences:

> The aim of sustainable development is to create a society where everyone has a good quality of life while maintaining and enhancing environmental resources.
>
> Chichester District Wellbeing Programme has something for everyone. (Chichester District Council, emphasis added)
>
> We must ensure everyone benefits from the new prosperity this Corporate Plan provides each of us with a direction of travel. And if we all pull in the same direction, we really can make Leicester more attractive for us all. (Leicester District Council, emphasis added)

Interestingly enough, a significant change has occurred over time in the level of the textual communication level by the city of Exeter. The local authority seems to fade away as a subject, leaving space for an operational presentation of the city that is *prosperous*, *learning*, *accessible*, *safe*, and *a city of culture*. In this sense, the changes made to the latest version of the text are particularly significant if compared to the previous ones. "A city where people are healthy and active" becomes "a city that is healthy and active"; "a city where the environment is cared for" becomes "a city that cares for the environment", "a city where everyone has a home" becomes "a city with homes for everyone" (emphasis added). Therefore, the (ideal) city is mostly represented as an active entity, as performing an active role, or better, personified. In fact, the city has become the meeting point of institutions and citizens' interests and needs.

4.2 Verbs

In line with the communication strategies outlined above, the use of the active voice is far more common than the passive. This stylistic choice avoids any impersonal or indirect connotation when it comes to describing the initiatives undertaken:

> The council supports local businesses in a number of ways.
>
> The service successfully runs throughout the District.

> The Council plays a key role within the community, from offering community grants to helping to tackle anti-social behaviour. (Chichester District Council)

On the one hand, the institution announces a policy and, on the other, it 'invites' the community to agree with it by calling upon that community, the very people. Active verbal forms help construct a set of responsibilities for local authorities, but also for citizens, who are called upon to take individual and collective action. This textual strategy depicts responsibilities for success and failure as shared, and this contributes to building consensus by assuming that there is no difference between the interests and values of the population and those of the institution:

> This is to assure us that the panel is as representative of the district as possible and that we get a wide range of people from the community. If you haven't received an invite to be on the panel you can still take part in other local consultation.

The community is thus seen as the true actor capable of influencing policies on the matter.

Such "empowerment" of community groups is expressed throughout the texts, making use of verbal forms such as "will give", "will enable", and "enabled":

> [we have] enabled more service users and carers to be involved in planning services. (Leicester City Council)

The use of the simple present tense has a dual function. It implies a categorical commitment of the producer to the truth of the proposition (Fairclough 1992), and this in turn supports a transparent perspective of the world, where facts are reported as categorical truths:

> Exeter has much to be proud of in terms of its rich heritage as well as well-established arts, civic events and community-based activities.
>
> Exeter is a great place to live and its people are proud of their city and are enthusiastic and energetic in their contributions to community life.
>
> Exeter is a prosperous city and people want to keep it that way. [...]
>
> This provides the basis for building an inclusive city where everyone has a role to play and is encouraged to take part. (Exeter City Council)
>
> It[1]s a challenge, but we have a roadmap to follow: this Corporate Plan provides each of us with a direction of travel.
>
> We want to try to make life better for every man, woman and child in a more welcoming city. (Leicester City Council)

Values focus on the present and future realization of outstanding performance, individual achievement, and community improvement. The modal verb *will* is used to highlight the strong desire to implement new future solutions to problems. This tense also serves a promotional purpose, as it is aimed at increasing the institution's credibility and creating consensus, as the following excerpts show:

> To achieve our strategic objectives we will…

> […] we will improve our services over the next five years to achieve our ambitious objectives. (Chichester District Council)

> Better communication will help ensure that our services meet our citizens' needs. (Leicester City Council)

> In many areas our priorities will be achieved through working in partnership with others […] to deliver the very best we can for Norwich. (Norwich City Council)

Another modal verb that seems to occur frequently is *can*. As it is most frequently used, *can* seems to indicate ability (which we infer to suggest possibility) as a result of the roles the participants assume, projecting a cooperative, collaborative stance, as the following excerpts suggest:

> Here you <u>can</u> discuss common issues and discover new solutions relevant to your particular business area. (Chichester District Council, emphasis added)

> But we <u>can't</u> make real and sustained progress on our own. Only by working with our partners and service users can we address the issues.

> And if we all pull in the same direction, we really <u>can</u> make Leicester more attractive for us all. (Leicester City Council, emphasis added)

> In many areas our priorities will be achieved through working in partnership with others […] to deliver the very best we <u>can</u> for Norwich. (Norwich City Council, emphasis added)

The promotional message conveyed by nominal and pronoun references, and by verbal structures and tenses is also reinforced by the use of predicates, which are typically associated with the council in the role of an actor. The most salient feature is represented by the choice of verbs belonging to the general semantic area of "improvement", such as *develop*, *innovate*, *regenerate*, *enhance*, *promote*, *build*, *cultivate*, *achieve*, and so on:

> The strength of the City Centre lies in its ability to support a wide range of retail uses […] that help to enhance the character and vibrancy of the centre. (Exeter City Council)

On the one hand, reference to the need to improve suggests that something was not adequate in the past.

> Many milestones and targets set three years ago <u>have been achieved</u>. (Leicester City Council)

On the other hand, public institutions aim to highlight the continuity of positive performance and achievements reached from the beginning of their mandate. On a linguistic level, this is realized through the use of presupposition, by iterative verbs (Levinson 1983) such as *continue*, *maintain* and *remain*, as seen in the following examples:

> We remain committed to our overall aim…
>
> Our commitment to partnership working is stronger than ever, as we strive to maintain the path of continuous improvement we embarked upon
>
> (Leicester City Council, emphasis added)
>
> The City Council will continue to work with identified partners in provision to ensure that delivery of those key items continues (Exeter City Council, emphasis added)
>
> We will continue to argue for a 'fair deal' for Norwich. (Norwich City Council, emphasis added)

Presupposition clarifies the validity of the implied statement, i.e., a standing level of good performance that is presented as shared opinion. Verbs expressing belief and commitment, such as *consider*, *believe*, *want*, *aim*, as well as *create*, *ensure*, *improve*, *maximize*, *minimize* are used to convey authoritativeness and a willingness to take on responsibilities, but also accountability for results:

> The team coordinates all our corporate consultation and offers advice and guidance to services to maximise the benefits of consultation. (Chichester District Council, emphasis added)
>
> Where there is a shared enforcement role or interest, either with other parts of the City Council, or with other agencies, we will work together to ensure consistency and a coordinated approach to make best use of available resources.
>
> […] we expect to meet the challenges […] in a way that minimises the impact on the environment and maximises the quality of life. (Norwich City Council, emphasis added)
>
> We want to remain a 4 star council, continue to improve services and provide value for money to citizens. (Leicester City Council, emphasis added)

Actions that need to be taken in order to achieve results usually come in the form of a set of 'priority areas' or 'corporate strategy pillars', making use of a typical communicative strategy in business. This is the case in Chichester and Leicester, where their vision statements are broken down into priority themes:

- Balancing the Local Housing Market
  We consider decent housing to be a basic and fundamental requirement in today's society, and we recognise the relationships which exist between housing, personal health, education, employment and family stability[.] (Chichester District Council)

- Improve quality and equality in teaching and learning
  Support children and parents, especially protecting the most vulnerable children. (Leicester City Council)

Imperative forms, which express advice, suggestions and recommendations, are used in the texts to encourage citizens to contribute to the improvement of their cities, thus reinforcing the dialogic component, as is visible in the following examples:

> Please see the study in the related content section for more information.
>
> Just click on the link to view the most recent edition.
>
> Have your say!
>
> To read the most recent Annual Letter for Chichester District Council [...] please view the related document. (Chichester District Council, emphasis added)

4.3 Nouns and Adjectives

The key verbs mentioned above (e.g. *provide, ensure, increase, encourage, develop*) are almost always associated with adjectives such as *best*, *efficient*, *effective*, *good*, *value-for-money*. This language serves a strategic, promotional function and stresses the agent's active role.

The following example epitomizes the increased importance that is attached to achievement and top performance, which are placed at the same level as the traditional concern for addressing citizens' needs by delivering quality services:

> The fastest growing economy in the east of England, it is home to the headquarters of 50 major companies, is one of the top shopping destinations in the country, and is the regional cultural capital.
>
> Growth will be used to bring benefits to local people, especially those in deprived communities, to regenerate communities, local economies, [...] by creating safe, healthy, prosperous, sustainable and inclusive communities. (Norwich City Council, emphasis added)

The frequent use of implicitly or explicitly positive terms such as *development*, *renewal* and *regeneration*, often related to health, education, training, and work, as well as adjectives like *welcoming, healthy, active, prosperous, attractive, vibrant* to describe the city represent an obvious reference to ideal living conditions for citizens:

> Attractive, safe and accessible parks and green spaces contribute to positive social, economic and environmental benefits improving public health, well-being and quality of life. (Chichester District Council)

Moreover, comparatives (e.g. better, higher, brighter) used in regard to the future of the city underline the institution's commitment to provide services that can help improve quality of life.

One final but important observation to be made relates to the use of the website as a medium of communication. Despite the fact that public communication, being institutional, must have a primarily informational purpose, and therefore refers to the rational sphere of its recipients, great emphasis is also put on the visual presentation of the place.

Figure 1: Images displayed on Leicester City Council website.

In the case of the figure above, the active role played by the institution (in this case, with reference to jobs opportunities and childcare facilities) is highlighted by being represented by its citizens. Images perform a strong emotional and evocative function and are not only meant to make the city more tangible, but also help underpin the logical arguments conveyed by the written text by enhancing, through the association of ideas, the interpretation process of the message.

5. Conclusions

This analysis exemplifies and clarifies the intention – expressed in such acts – of institutional communication to build a society which, if not as perfect as utopian models, should at least be a better one. It should be a society able to recognize the needs of present and future generations: sustainable development, economic prosperity, cultural vitality, social cohesion.

The following excerpt from Norwich City Council website seems to best summarize the concept:

- Everyone will have access to suitable housing that reflects their needs.
- People will enjoy healthy, safe and fulfilling lifestyles, have equitable access to high standards of health and social care and make informed choices about their own health.
- There will be excellent opportunities for lifelong learning and personal development and people will have high expectations for their own educational achievement to meet their needs, to contribute to the life of their communities, and to the economy.
- The area will be renowned for its culture, creativity and spirituality, with high quality cultural and leisure opportunities that improve people's well-being.

The reformed public sector struggles to combine "partnership" with competition, giving shape to an alternative perspective of publicness in which the system from the "public" becomes of "public utility", and the state is no longer a "subject" but a "function".

By using technological tools as facilitators in inclusion processes, this contemporary utopia is therefore an e-utopia, the project of an ideal and idealized, but also desirable and possible, democracy. It thus becomes an e-democracy where the utopia wants and manages, or may be able to be attained.

References

Bait, M. (2008). "Language and Identity in Public Administration: A Case-Study of the City of Belfast". In: Di Martino, G., V. Polese, and M. Solly (eds) *Identity and Culture in English Domain-specific Discourse*. Naples: Edizioni Scientifiche Italiane: 281–296.

Bait, M. (2009). "The Representation of Mission on Institutional and Corporate Websites: A Case of Migration of Discursive Practices". In: Torretta, D., M. Dossena, and A. Sportelli (eds) *Forms of Migration. Migration of Forms*. Bari: Progedit, 401–416.

Benveniste, É. (1966). *Problèmes de linguistique générale I*. Paris: Gallimard.

Coleman, S. (2001). *Democracy Online*. London: Hansard Society.

De Fina, A. (1995). "Pronominal Choice, Identity, Solidarity". *Political Discourse. Text* 15(3): 379–410.

Fairclough N. (1995). *Critical discourse analysis: The critical study of language*. Longman: London.

Fairclough, N. (2001). *Language and Power*. Harlow: Pearson.

Featherstone, M. (1991). *Consumer, culture and postmodernism*. Sage: London.

Habermas J. (2006). "Contemporary Political Philosophy: An Anthology". In: Goodin, R., and P. E. Pettit (eds). 2nd ed., Malden, MA: Blackwell: 103–106.

Hackney, R. A. and K. McBride Neil (1995). "The Efficacy of Information Systems in the Public Sector. Issues of Context and Culture". *International Journal of Public Sector Management* 8(6): 17–29.

Halliday, M.A.K. (1994). *An Introduction to Functional Grammar*. London: Edward Arnold.

Jeffres, L. W. and A. C. Lin (2006). "Metropolitan Websites as Urban Communication". *Journal of Computer-Mediated Communication*, 11(4): 957–980.

Hunston, S. and G. Thompson (eds) (2001). *Evaluation in Text*. Oxford: Oxford University Press.

Karakaya, P. R. (2005). "The Internet and Political Participation. Exploring Explanatory". *Links European Journal of Communication*, 20(4): 435–459.

Terry, L. D. (1998). "Administrative leadership, Neo-Managerialism, and the Public Management Movement". *Public Administration Review*, 58(3): 194–200.

Villa, D. R. (1992). "Postmodernism and the public sphere". *American Political Science Review*, 86: 712–721.

West, Darrell M. (2004). "E-Government and the Transformations of Service Delivery and Citizen Attitudes". *Public Administration Review* 64(1): 15–27.

Wilhelm, A. G. (2000). *Democracy in the Digital Age*. New York and London: Routledge.

Wilson, J. (1990). *Politically Speaking. The Pragmatic Analysis of Political Language*. Cambridge, MA: Basil Blackwell.

Wouters, C. (1986). "Formalization and informalization: Changing tension balances in civilizing processes". *Theory, Culture and Society* 3(2): 1–18.

Federico Boni

The Utopia of Communication The Myth of Communication as a Positive Value

1. Introduction: Communication, Utopia and Myth

This chapter aims to investigate a particular kind of utopian communication: the idea of communication as a positive value. In order to understand how communication can be framed within the context of utopia, it is important to remember that the history of communication itself (interpersonal, mediated and mass-mediated communication) is full of cultural, social and political utopias and, in particular, technological utopias. Communication studies have been characterized by the traditional binaries of communication versus miscommunication, and utopia versus dystopia, which are embedded within communication theories of the nineteenth and twentieth centuries (from the social constructivists to the pragmatists; from information theory to the psychological models).

Further, it is important to understand exactly what is meant when we talk about communication, utopia and myth.

Broadly speaking, there are two types of definitions of communication. The first sees communication as a process by which A sends a message to B; accordingly, this message has an effect on B. The second type of definition sees communication as a negotiation and exchange of meaning; messages, people and society interact to produce meaning. While the aim of the first type of definition is to identify the stages through which communication passes so that each one may be properly identified and studied, the second type focuses on the relationship between constituent elements required for meaning to be produced within a cultural and social context.

From a sociological point of view, communication is a very particular kind of social interaction. The most comprehensive typology of the main forms of interaction was proposed by John Thompson (1995). According to his model, the main three types of interaction are face-to-face, mediated and mediated quasi-interaction. In face-to-face interaction, the actors involved in the interaction are present, which affords the possibility of using a variety of symbolic cues. Under mediated interaction, we have communicative processes such as letter writing or telephone conversations, and of course all the types of computer-mediated communication. The defining characteristic of mediated interaction is the use of a technical medium, which narrows the availability of symbolic cues and makes individuals more dependent on their own interpretations and contexts. Finally, the mediated quasi-interaction is the communicative process established by the traditional mass media, based on a monological interaction.

For our analysis of the utopia of communication, a useful point of reference is Karl Mannheim's (1976) classic sociology of knowledge. His emphasis on the 'concreteness' of utopias means that utopian thinking is considered as an active social force affecting the transformation of societies. Mannheim frames his concept of utopia within the theoretical framework of the sociology of knowledge because he argues (indeed, in a very sociological fashion) that the history of ideas should be seen as the history of their uses. Thus, the argument is that utopias do not impact history as abstract ideas but rather have an effect through the concrete use of those ideas by individuals, social movements, institutions, etc.

Far from an everyday and commonplace understanding of utopian thinking (where utopia is seen as wishful thinking), Mannheim turns utopia into a situational and socially conditional concept. As we will see, utopian (and dystopian) thinking about communication are conditional at least at a discursive level, given that we define 'discourse' as 'language in use', that is, language used in some context, for some purpose. If discourse is what we create when we use language in social contexts, it is possible to see how a linguist's definition of discourse as 'language in use' might relate to a social theorist's definition of discourse as 'a social construction of reality', a 'form of knowledge'. In the end, this whole process is actually a communicative act, a kind of performative 'speech act,' and in this way, communication becomes the method through which we can construct utopian (or dystopian) discourses about communication itself in a self-reflexive process of social construction of reality.

By analyzing discourses, we are able to argue that certain ways of representing communication processes have become 'naturalized', so that people no longer recognize them as incorporating a political or ideological stance. The discursive pattern is a clue to what is taken as simple common sense on a particular issue (in our case, communication), and the repetition of the pattern means that it will continue to be common sense.

This leads us to the concept of 'myth'. Our use of this concept is far from the anthropological meaning, where 'myth' refers to a narrative that offers explanations of why the world is as it appears to be, and why people act as they do. Our use of the concept is closer to the semiotic understanding of the term, where 'myth' refers to an unarticulated chain of associated concepts and discourses by which members of a society understand certain topics. Its prime function is to make the cultural natural, which is a distinctive feature of ideological discourses. The ideological productivity of naturalization is that circumstances and meanings that are socially, historically, economically and culturally determined (and hence open to change, as they are the result of signifying struggles) are experienced as natural, that is, inevitable, timeless, necessary, unarguable (Barthes 1973).

The concepts of discourse and myth are useful tools because they help us be attentive to the distinct uses of language and communication. As discourses and myths, utopias are contextual and situational and are therefore expressed and used in a number of different articulations. In the struggle between utopians and representatives of the existing order, those articulations are contested, and differing, even contradictory interpretations are produced.

2. The Utopia of Redemption

In his book *L'Utopie de la communication*, Philippe Breton (1992) traces the origin and development of the notion of the utopia of communication to the growth of cybernetics during and after World War II. Breton argues that the idea of communication is not modern, as it has existed for centuries. What is new and truly modern, however, is the manner in which it has expanded to encompass the entire intellectual landscape by the end of the twentieth century and at the beginning of the twenty-first.

Born of various attempts to formulate theories of command and control in weapons systems in the effort to win World War II, cybernetics provoked an incredible theoretical explosion that led to speculation in fields soon removed from the context of war. Breton argues that the computer was a counterpoint of the atomic bomb, a kind of moral redemption of the bomb: the theorists of information theory and cybernetics were the same people who worked on the creation of the deadly weapon. The information age, which was in force during the second half of the twentieth century, was marked by a struggle for redemption and purification between Good (the new utopia of the democratization of communication brought by the 'information age') and Evil (the world of war where this utopia was born).

Nevertheless, the relationship between war and communication persists, especially in the age of electronic mass communication. The structural relationship between war and media presents forms and patterns that vary depending on the kinds of conflict and actors involved. A common theoretical assumption is that the wars that took place in the media era represent the failure of McLuhan's prediction. Foreseen more than 40 years ago, McLuhan argued that since the war in Vietnam, we have entered the age of the "TV war", where the audience, from their homes, "participate" in each stage of the conflict (McLuhan and Fiore 1968). On the contrary, we can argue that war, in the age of media representation, is *negated* through the use of linguistic and discursive rhetoric; furthermore, the very representation of the conflict is heavily censored, with restriction – and sometimes the impossibility – of its visibility.

Another theory is about the structural relationship between war and media, where this relation must be understood as a flow of information on – and a representation of – the conflict, but also as an organic connection between the media system and the political and military powers. Thus, the media can be understood as a means of communication between politicians and citizens, but also between politicians and diplomats, and between military strategy and public opinion (the so-called 'media diplomacy', see Cumings 1993).

From the point of view of the media discourse strategies, another theory is the 'mediatization of war', a concept which has to be defined and analytically explored. The concept of the 'mediatization of war' refers not only to the relationships among war, the media and political and military systems that we have seen before, but also uses media languages and codes to define and frame war (Malek and Kavoori 1999). In this sense, it is necessary to include in the definition of media the modern Information and Communication Technologies (ICT), which account for the contemporary communication technologies more accurately and realistically. As a matter of fact, information technologies play an important role in the very emergence of the forms contemporary conflicts take. The cycle of war is fulfilled within an IT logic, from remote sensing (surveying territory from the satellite) to 'photo finishing' (photos taken after a raid). This 'visual' nature can be seen within media discourse strategies, which tend to minimize the prices of war in terms of human lives, through an 'aseptic' representation of conflict. It is aseptic, and thus surgical, just like the very definition of wartime activity in Western countries, by using representational strategies that are precise and abstract.

Abstraction itself, along with stylization, is an element of the mediatization we are defining: according to an established media logic, the rhetoric of conflict focuses on some image-symbols, which sterilize and stylize the war. This is certainly a targeted discursive strategy, but this is also part of an established routine practice, typical of media language. An important aspect of the mediatization of war is that it is connected to concepts of 'personalization' and 'narrativization'. Personalization concerns not only the enemy, but also the victims. This can be seen in the paradigmatic individual stories that give a name and a face to the masses of refugees shown on TV. Again, the dramatic narrative formula that amplifies individual stories belongs to the media logic used in 'reality television'.

Finally, the concept of the mediatization of war acknowledges the role played by the new media, mostly by the Internet, in the new wars. The theoretical importance of the entrance of new media into the relationship between media and war is obvious with regard to the possibilities of communication and information (either interactive or not), but mostly with reference to the agenda-setting process (McCombs et al. 1997). Hypertextual information sources maintain few to none of the typical forms of presentation used in traditional media, leaving the user with the task of selecting and hierarchically ordering news and information.

3. The Utopia of Communication as Democracy

In Breton's analysis, the social theory that emerged from communication theory can be seen as an ideology that clearly reiterates some of the principal elements characterizing the utopian thought of nineteenth-century socialist thinkers. From the notion of small communities to that of the absence of social hierarchies, communication theory in its form as social philosophy and social theory puts a modern face on old utopias.

Further, as mentioned in the previous section, the very thinkers who championed cybernetics and communication as a means to win the most dreadful war in human history – people like Norbert Wiener – then turned around and presented communication as a means to end all wars. It is paradoxical that the very means of contemporary communication are so closely connected to warfare, as we saw in the previous sections. The transparency of human relations that communication supposedly fostered would bring a permanent end to the possibility that the dark secrets of genocide could be reproduced in the shadows, hidden from the sight of society at large and shielded from public opinion.

The utopia of transparency of communication is one of the most enduring discourses and myths in our contemporary age. Electronic and digital media (television and the Internet) are seen as 'windows', an image that relates the idea of mass and personal communication to the idea of openness and transparency. Endemol's *Big Brother* is one of the most interesting metaphors depicting television's openness towards the world, where we can find the utopian element of television as an open (reversed) window and the dystopian element as 'prison as entertainment', which involves the deliberate sequestration of participants in contained spaces. In fact, *Big Brother* essentially mirrors the contemporary social experiment in which neo-liberal economic doctrine is extended indiscriminately into the fabric of intimate life. When the self is experienced as a media commodity, the modern project of 'openness' loses its way, and the dream of transparency is turned into a nightmare.

After all, the idea of communication as democratization can be linked to classic political liberalism, its conception of individual freedom and the idea of public sphere. In this sense, the mythical utopia of communication as the process of democratization represents the reinterpretation of liberalism in the context of the information society, media society, and, later, network society. The democratic utopia of communication constructs a vision of a society in which new media and all new forms of communication technology would open up a way towards the 'polis', the democratic community of equal individuals. These new information and communication technologies would facilitate the direct participation of people in political and social debate, as well as in decision-making, without the control of the bureaucratic state. This utopia is characterized by a nostalgic return to the concept of 'community', without realizing that the very concept of 'community' is at the same time an inclusive and exclusive one, including few and excluding many from the communicative process.

Technological changes in the possibilities of community in information societies have been accompanied by radical philosophical rethinking of the meaning of community and association (see Nancy 1991; Agamben 1993). The concept of a 'virtual community', like the very concept of utopia, is seen as inherently flawed if it is based on existing modes of interaction. Cameron Bailey argues that the vision of the Internet as an ideal democratic community in the mold of the Greek agora "contains its own ideological dead weight [...]. Like the democracy of the ancient Greeks, today's digital democracy is reserved for an elite with the means to enjoy it" (Bailey 1996: 31). We should ask whether or not the Internet possesses enough stability for the ongoing recognition of members of Internet 'communities' and the stability of their own sense of self. Further, it is quite difficult to refer to this new media scenery as the formation of 'communities' because it is based on a very vague acceptance of the word 'community', given that 'social networks' such as *MySpace* have over 100 million members.

The perverse effect of this typically utopian view is the systematic denial of conflict, which is demonized, reduced to the only dimension of violence.

Another discursive and mythical utopia of communication as a positive value is that of the 'networked society', which is linked to the idea of the 'web' and the 'net'.

The 'cult of the network' is historicized by Armand Mattelart in his *The Invention of Communication* (1996), in which the French scholar argues that the different utopias regarding communication 'invent' the very different concepts of 'communication', thus creating the discourses and the myths that frame both the theoretical and common-sense thinking about communicative processes.

In Mattelart's opinion, every technology involved in "the multiple circuits of exchange and circulation of goods, peoples, and messages" was a technology of communication (1996: 34). For example, the Saint-Simonian conception of a communication technology, the 'cult of the network' as Mattelart calls it, was broad enough to include a network of railroads and an advertising network, along with networks of journals, banks, and industrial fairs. In a very fascinating way, Mattelart argues that the contemporary rhetoric about a communication revolution was the ideology (and the utopia) of the whole of historical capitalism. Further, he makes a clear case about the depth and breadth of the pattern of ideologizing communication as an agent of social revolution.

Mattelart organizes his argumentation around four histories: (1) communication technology as producing social flow (rational fluidity / enlightened state administration; market fluidity / liberal political economy; evolutionary fluidity / Darwinian social theory); (2) place (world's fairs, Fourier's Phalanstery); (3) space (national and imperial, linguistic and cultural, religious and military); and (4) norm (of a psychological and physiological social individual, of a market consumer). Mattelart employs a definition of communication technology that includes the whole circuit of exchange, adding the exchange of materials and people to the exchange of signs and data. By employing this historical perspective, he makes the argument that this utopic ideology has been a force in recent centuries, not just in recent decades.

Nevertheless, it is true that the recent democratization of access to Internet tools promotes a significant intensification and acceleration of the trend of circulating exchanges, transactions and practices of creating or adapting-appropriating existing cultural content to redistribute this content once it has been transformed. Mass communication assumes a new meaning in this context, as new forms of communication are emerging: alongside interactions between individuals (telephone and email) and a growing amount of group communication (chat, forums and discussion lists), new patterns are emerging. Thanks to new social media, individuals can achieve instant visibility and can speak to the masses on a global scale. These phenomena become particularly visible during terrorist attacks or natural catastrophes: bloggers are the first to broadcast the first information and photos (taken with their mobile telephones) from the scene of the event. Moreover, we are witnessing the intrusion of amateurs into the world of professional journalism. Web 2.0 ushered in an era of communication of the masses, but also *communication by the masses, for the masses.* In light of this trend, the view of the majority assumes new importance regarding the authority of experts. Some media elites may be shaken by this explosive informational trend.

All this has made it possible to reach a new level in the paradoxical realization of the utopia of universal access to knowledge formulated by some of the Internet's founding fathers. This utopia is paradoxical because this explosion of knowledge in the spread, distribution and creation of information (e.g. texts, photos, music and videos) may, at the same time, induce a feeling of incompleteness on the part of heavy users, with its invitation to the frantic and infinite search for information that is constantly being updated, and to which information can always be added, creating an infinite, ongoing search without end.

4. The Utopia of Effective Communication

There is another way in which the communication made possible by new media is paradoxical: with Web 2.0, many scholars observe the return to the 'old' patterns of interpersonal communication in the form of 'word of mouth'.

Word of mouth is a 'pass-along' process of interpersonal communication, a 'social diffusion' that has particularly powerful effects, especially in its ability to influence people as messages are 'passed' and transmitted from person to person. Based on the traditional model of the 'two-step flow of communication', under which media messages are mediated through interpersonal relationships, word of mouth includes face-to-face discussions as well as so-called 'word of mouse' online interactions taking place on social media. Word of mouth can be considered a new utopia of 'effective communication' with which it is possible to achieve new kinds of communication possibilities by using a very old communication tool.

Broadly speaking, word of mouth is considered one of the most common and influential channels of communication. In general terms, it can be considered the main feature of interpersonal communication. The first channel of word of mouth is face-to-face interaction, but we can have word-of-mouth communicative processes with different technologies of communication (e.g. by telephone or computer). As such, word of mouth includes literal interpersonal communication, or face-to-face conversations and discussions, as well as mediated communication and the so-called 'word of mouse', that is, online interpersonal communication.

Word of mouth is a communicative process that focuses mainly on face-to-face and mediated interaction, but its importance stems from its ability to connect interpersonal communication to mass communication (Thompson 1995 – see Introduction).

In fact, since the beginnings of communication research, the most relevant studies and theories of mass communication have shown that studying the media inevitably entails studying the processes of word-of-mouth interpersonal communication. The works of Paul Lazarsfeld and Elihu Katz have clearly indicated the importance of interpersonal interaction, even if we want to investigate the social effects of the media. In an early study which focused on the analysis of the motivations and the ways in which political opinions are formed (in this case, the panel comprised six hundred voters of Erie County, Ohio, during the 1940 presidential campaign [Lazarsfeld *et al.* 1948]), the findings were quite interesting: the effectiveness of mass communication is deeply – and causally – linked to processes of interpersonal communication. Lazarsfeld speaks of a "two-step flow of communication" to illustrate the role of the mediation of "opinion leaders" (people with a good level of information, in direct contact with the media) between the media and other individuals of the public. The effects of the media are thus only a part of a broader process, namely personal influence. This assumption was taken up by the same author in a subsequent search, conducted by Katz, which focused on consumption (Katz and Lazarsfeld 1955). The findings confirm the hypothesis of the 'two-step flow of communication', showing that interpersonal communication has a greater degree of flexibility in the face of the resistances of the receiver compared to mass media. If, in a communication process, the credibility and the reliability of the source affect the influence of the message, then it is likely that the impersonal source of the media is at a disadvantage compared to the reliable sources of interpersonal relationships.

The role of interpersonal communication is also irreplaceable as a source of information, albeit not a primary one. In a study of how the news of the assassination of John Fitzgerald Kennedy reached the public, Bradley K. Greenberg (1964) showed that many people had been informed of the event not directly by the media (e.g. radio, TV, newspapers), but by friends and acquaintances (who, of course, had heard the news from the media). In more general terms, it would seem that face-to-face interaction is particularly important in acquiring both the most groundbreaking news, along with news more related to the local context or to specific personal interests.

Further research on word of mouth has shown that the main effects of this kind of communication impact individuals in two key ways: awareness (informing people that a behavior exists, or that a fact has occurred, as in the cases shown above) and persuasion (word of mouth can change opinions about whether something is right or worth doing, or a particular party or political leader is right or worth voting for). Word of mouth can also influence the social identity of individuals associated with a particular party or behavior, which may, in turn, affect the likelihood of interest or vote.

One of the main features of word of mouth is the source. People tend to listen to more credible and reliable sources, or those that come from sources considered to be trustworthy or to have expertise in a certain area, such as political issues. Thus, what should be stressed here is the importance of the strength of the ties (friendships or acquaintances, that is, strong or weak ties). On the one hand, strong ties may have a greater impact and stronger effects because people tend to trust them more, since they know more about their interests, tastes, lifestyles and behaviors. On the other hand, people generally tend to have a greater number of weak ties, or acquaintances, so the overall influence of these individuals may be stronger. While word of mouth from similar people may have a greater impact (because, for instance, their tastes and political views are similar), word of mouth from individuals who are less similar may offer access to different information and alternative views and perspectives.

Traditional research on word of mouth in political communication has focused on small groups (especially primary groups) and small communities in local contexts. A study conducted by Lenart (1997) investigated the spread of name recognition of candidates seeking the 1992 U.S. Democratic presidential nomination in the Iowa caucuses and the New Hampshire primary election. The study demonstrates the comparative influence of media exposure and interpersonal discussion. As the author argues, interpersonal communication depends on the logics of media coverage, in both complementing and strengthening media effects when they are strong, and substituting for media effects when they are nonexistent.

In discussing the role of the mass media and interpersonal discussion in local political participation, McLeod, Scheufele and Moy (1999) focused their study on community integration in mass and word-of-mouth communication, showing that viewing television news had no direct impact on political participation, but it did have a certain effect on more institutional and conventional forms of participation. A few years after, a study conducted by Scheufele (2002) focused on the idea that interpersonal discussion among citizens is the 'soul of democracy' (an idea that, as Scheufele argues, has been treated almost as a truism in mass media and interpersonal communication research), showing that while it is generally true that interpersonal discussion of politics is a key precursor of political participation, the relationship between hard news media consumption, interpersonal discussion of politics and political participation is a complex process. While consuming hard news media has a great influence on political participation, this effect is more evident for people who talk to others about politics than for those who do not. Scheufele et al. (2004) focused on different settings for primary discussion networks (the workplace, church and volunteer groups) and their effects on political participation, showing that the social setting where citizens discuss politics is an important predictor of political participation. For example, discussion networks in volunteer groups had a great impact on recruitment, thus demonstrating that discussing politics frequently in this setting is directly (and positively) linked to political activity. Conversely, the impact of conversational networks in church and workplace settings on participation was only indirect. Another study focusing on word-of-mouth communication, media exposure and political participation (Sotirovic and McLeod 2001) examined how communication patterns mediate the influences of values on political participation. According to the authors, individuals' efforts to think about news and search for additional information and perspectives has a modifying effect on what people get from the media. The more people integrate information from various sources, the better they understand the political world, which ultimately has a positive effect on political communication.

In this vein, scholars argue that powerful political campaigns are not only those that inform and persuade, but also those that motivate further interest and talk, thus confirming Carl and Duck's (2004) previous conclusion that "a person is more likely to be influenced by messages that stimulate discussion between the individual and his or her groups of significant others or associates" (26). Voters use oral communication as a tool while making a decision and getting opinions from their social circles.

With the advent of social media, word-of-mouth communication has gained more attention. The 'online' version of word of mouth, known as 'word of mouse', includes all new channels of the digital media, such as chat rooms, blogs, newsgroups, *Facebook*, *Twitter* and social media websites in general. Research on word of mouse differs from traditional research on word of mouth because electronic word of mouth can be easily accessed, linked, and searched. Compared to traditional word of mouth, online word of mouth has the potential to be far more influential and make a greater impact due to its speed, convenience, wide reach, and the absence of face-to-face pressure and intimacy. Furthermore, by using search engines such as *Google*, it is possible to seek out the opinions of strangers, an opportunity which is quite rare in traditional word of mouth processes. Finally, electronic word of mouth provides opinion leaders with a more efficient means of disseminating information, and facilitates the searching of information for opinion seekers.

The emergence of the Internet and of online social networks has led to a self-organizing propagation process that recalls the dynamics of an epidemic, a 'cascade of influence'. On the one hand, this new form of the 'information cascade' allows citizens to voice their opinions, and it mobilizes communities and voters around their candidates. On the other hand, new participative media have changed the way elections take place by allowing politicians to reach new audiences with new 'viral' forms of political marketing. In addition to this, electronic word of mouth is likely to change collective and political behavior by transforming traditional 'old' social networks, which are normally limited in size and scope, into networks potentially composed of millions of individuals.

5. Conclusions

In the preceding sections, we have seen that the 'dialogical' model of communication – that is, the model of interpersonal communication – is considered the most effective form of communication, adding positive value to communicative processes. To move towards a conclusion, it is worth arguing that this isn't always true. According to some scholars, the myth of 'good communication' is attainable through the old model of mass-mediated communication, the 'one-way' model. In the sensible link of thought, 'good communication' (communication as a positive value) is based on the dialogical model: there must be no secrets or locked doors between communicators. Each should be fully, genuinely and sincerely open to the other.

In his book *Speaking into the Air* (1999), John Durham Peters identifies two great models of communication: dialogue and dissemination. Each is a principle and a practice, and their exemplary practitioners were Socrates and Jesus. Socrates' model is a 'love model' of communication, a love connection between two people, both of which are alive and present to each other. The ideal (utopian) human relationship is the fusion of both. Socrates, Peters tells us, argued that insemination is more virtuous than dissemination. Insemination is to implant the seed in another where it will bear fruit. Dissemination is like the sin of Onan who spilled his seed upon the ground. It is wasteful to scatter seed, for there is no guarantee that the seed will bear fruit. Cast in these terms, Christ's method of communication appears scandalously inefficient.

Jesus's discourse and methods stand in sharp contrast to those of Socrates. Both are exemplified in the parable of *The Sower*; a story with a message told to a large crowd on the shore of the Sea of Galilee. Instead of the Socratic one-to-one dialogue, a form of two-way communication, we have one-way communication between a single speaker and an anonymous mass of listeners. The story of the sower makes explicit the significance of communication as mass dissemination or broadcasting. The sower in the parable scatters his seed indiscriminately. Some, as Jesus tells it, fell on stony ground and were picked up by the birds of the air. Some fell among thorns and were destroyed. Some fell on shallow soil and grew quickly but soon withered and died. And some fell on fertile ground and yielded a good harvest; thirtyfold, sixtyfold, a hundredfold. It is, of course, a parable about parables – Jesus's own account of his way of spreading the Word.

Peters, then, offers two paradigms of communication; one a dialogue of intimacy and reciprocity, the other of indiscriminate mass dissemination. It is a contrast between the personal and impersonal, individual and social, present and absent, embodied and disembodied relationships. Today, we generally take the intimate paradigm as the norm and see the impersonal paradigm as deviating from the mark. But it is rather clear that Peters prefers non-reciprocal, one-way communication. And he shows – quite convincingly – that the idea of 'communication' (the *invention* of communication, as Mattelart would put it) emerged from a late Victorian cultural milieu combining spiritual mysticism, scientific experimentation, and popular fascination with communicating with souls and angels, and exchanging messages with the dead. Modern modes of communication – photographs, phonographs, telephones, radios –resulted in "a new kind of quasi-physical connection across the obstacles of time and space" (5). In a culture fascinated with paranormal phenomena, these technologies produced new sites for the ghostly presence of disembodied others, and inspired reflection on the mysterious channels of communication. Peters reveals the irony surrounding new technologies of communication that reactivated primal doubts about the division of the mind and body and chasms in communication.

The spectral utopia of communicating with the dead is a common feature of all the technologies of communications of the nineteenth and twentieth centuries. To be sure, communication and media technologies became 'haunted' as soon as they were introduced: as Jeffrey Sconce (2000) argues, a fascination with communication developed in both the occult and scientific fields as soon as the new communication technologies appeared. The telegraph merged 'electromagnetic' and 'spiritual' communication, thus connecting technology with spiritualism and the occult. Alexander Graham Bell's assistant was also a medium, and Bell himself attended séances; Thomas Edison and Guglielmo Marconi both conceived of the theoretical foundations for their devices as ways to contact the dead. And if radio broadcasts appeared as a mysterious 'voice from the void', a way to contact ghosts via wireless, the 'television ghosts' created by the eerie images appearing on a TV set during signal interference seemed to bring the spectral world into the home, rendering television the quintessential 'haunted medium'. The technological 'medium' shares the uncanny qualities of the human 'mediums', connecting the world of the living with the world of the dead, and the prefix 'tele-', used in so many electronic media devices including the television itself, is also attached to 'telepathy' and other paranormal phenomena such as 'telekinesis' and 'teleplasm' (Warner 2006).

Interestingly enough, the television as the uncanny medium is strictly connected with the dystopian and dysphorian Gothic trope of the haunted house.

After all, television is a mysterious device which is simultaneously inhabited by ghostly images of ourselves and it inhabits our homes (Lewdon 1993). If television is "a profoundly domestic phenomenon" (Ellis 1982: 113), then it can create a sense of the uncanny by introducing the unfamiliar to the familiar, thus bringing the uncanny (*unheimlich*) into the familiar (*heimlich*) of the home (Lewdon 1993). Just as the *heimlich* contains the *unheimlich*, so too does the domestic environment of the home contain the paranormal potential of television. Television is "the ghost in the home" (Lewdon 1993: 70), a communication device which serves as an access point for horror to enter the home (Wheatley 2006), whose very presence in the living room becomes a metaphor for the anxieties and paranoias of the domestic space and family life. Thus, if we consider the Gothic trope of the 'haunted house' and the Gothic concerns for the electronic media of communication, we have two different houses, the first being that of one of the most domestic genres (Gothic) and the second being that inhabited by the ghostly and paranormal presence of the television set (Wheatley 2006).

At the same time, we have two different kinds of uncanny presences within the familiar space of the home: the presence *of* television, and the presence *in* television, which add a paranormal and ghostly layer of meaning to television's quality of 'liveness'.

The fear of the 'live' quality of television is connected to the 'live presence' of the medium in the home of the viewer; this uncanny presence turns 'liveness' into a ghostly and spectral 'deathness' which links the television set and content to the *unheimlich*, the 'unfamiliar'. Television's 'liveness' assures that what we are watching is happening right now, thus enhancing the utopian illusion that what is being shown is 'real'. It is important to stress here that 'liveness' is seen by many media scholars as an ideology, a utopia, an argument that, as we have seen, can be extended to the claim that television shows us 'reality' (Feuer 1983). That of 'liveness' is not only a ritual category, as it "guarantees a potential connection to our shared social realities as they are happening" (Couldry 2003: 96–97), but it is probably the longest-standing myth of television's ontological presence and essence. Vianello (1985) argues that 'live television' is a strategy of business practice and domination, a "power politics" which, in spite of the old fears of "Deathness", is "still alive" (Bourdon 2000) – perhaps as a specter haunting our television sets and our everyday lives.

This dystopian scenario of a very ghostly kind of communication represents the concerns and the anxieties of almost 150 years of electronic communication and nearly 100 years of scholarship on the effects of media. Popular representations of 'media *presence*' offer a fascinating counterpart to the established scholarship on media and communication studies, often letting the ghost of repressed issues haunt the spaces and the times of the media and of our lives. As we have seen, this dystopian and dysphorian scenario stems from the utopia of communicating with the dead in another dimension. This is the 'dark side' of the paradoxical quality of all communication utopias, which began at least with the myth of the Tower of Babel, simultaneously the symbol of a utopia of communication and of the disastrous consequences that followed when attempting to build that very utopia.

References

Agamben, G. (1993). *The Coming Community*. Minneapolis: University of Minnesota Press.

Bailey, C. (1996). "Virtual Skin: Articulating Race in Cyberspace". In: Moser, M.A and D. MacLeod (eds) *Immersed in Technology: Art and Virtual Environments*. Cambridge, MA: MIT Press, 29–49.

Barthes, R. (1973). *Mythologies*. St. Albans: Paladin.

Breton, Ph. (1992). *L'Utopie de la communication*. Paris: Gallimard.

Bourdon, J. (2000). "Life television is still alive: on television as an unfulfilled promise". *Media, Culture & Society* 22(5): 531–556.

Carl, W. J. and S. Duck (2004). "How to do things with relationships and how relationships do things with us". *Communication Yearbook* 28. New York: Lawrence Erlbaum, 1–36.

Couldry, N. (2003). *Media Rituals. A critical approach*. London and New York. Routledge.

Cumings, B. (1993). *War and television*. London and New York: Verso.

Ellis, J. (1982). *Visible Fictions: Cinema, Television, Video*. London and New York: Routledge.

Feuer, J. (1983). "The Concept of Live Television". In: Kaplan, E. (ed.) *Regarding Television*. Los Angeles: American Film Institute: 12–22.

Greenberg, B.S. (1964). "Person-to-Person Communication in the Diffusion of News Events". *Journalism Quarterly* 41: 489–494.

Katz, E. and P. F. Lazarsfeld (1955). *Personal Influence: The Part Played by People in the Flow of Mass Communication*. Glencoe, IL: The Free Press.

Lazarsfeld, P. F., B. Berelson, and H. Gaudet (1948). *The People's Choice: How the Voter Makes up his Mind in a Presidential Campaign*. New York: Columbia University Press.

Ledwon, L. (1993). "*Twin Peaks* and the Television Gothic". *Literature/Film Quarterly* 21(4): 260–270.

Lenart, S. (1997). "Naming names in a Midwestern town: The salience of Democratic presidential hopefuls in early 1992". *Political Behavior* 19: 365–382.

Malek, A. and A. P. Kavoori (eds) (1999). *The Global Dynamics of News: Studies in International News Coverage and News Agenda*. Stamford, CT: Jai Press.

Mannheim, K. (1976). *Ideology and Utopia: An Introduction to the Sociology of Knowledge*. London: Routledge & Kegan Paul.

Mattelart, A. (1996). *The Invention of Communication*. Minneapolis: University of Minnesota Press.

McCombs, M., D. L. Shaw, and D. Weaver (1997). *Communication and democracy: Exploring the intellectual frontiers in agenda-setting theory*. Mahwah, NJ: Erlbaum.

McLeod, J. M., D. A. Scheufele, and P. Moy (1999). "Community, Communication, and Participation: The Role of Mass Media and Interpersonal Discussion in Local Political Participation". *Political Communication* 16: 315–336.

McLuhan, M. and Q. Fiore (1968). *War and Peace in the Global Village*. London and New York: Random House.

Nancy, J. L. (1991). *The Inoperative Community*. Minneapolis: University of Minnesota Press.

Peters, J. D. (1999). *Speaking into the Air. A History of the Idea of Communication*. Chicago: University of Chicago Press.

Scheufele, D. A. (2002). "Examining Differential Gains from Mass Media and their Implications for Participatory Behavior". *Communication Research* 29: 46–65.

Scheufele, D. A., M. C. Nisbet, D. Brossard, and E. C. Nisbet (2004). "Social Structure and Citizenship: Examining the Impacts of Social Setting, Network Heterogeneity, and Informational Variables on Political Participation". *Political Communication* 21: 315–338.

Sconce, G. (2000). *Haunted Media: Electronic Presence from Telegraph to Television*. Durham, NC: Duke University Press.

Sotirovic, M. and J. M. McLeod (2001). "Values, Communication Behavior, and Political Participation". *Political Communication* 18: 273–300.

Thompson, J. B. (1995). *The Media and Modernity. A Social Theory of the Media*. Cambridge: Polity Press.

Vianello, R. (1985). "The power politics of 'live' television". *Journal of Film and Video* 37(3): 26–40.

Warner, M. (2006). *Phantasmagoria: Spirit Visions, Metaphors, and Media into Twenty-first Century*. Oxford: Oxford University Press.

Wheatley, H. (2006). *Gothic Television*. Manchester: Manchester University Press.

Paola Bozzi

Sex and the City: Berlin and the Utopia of a New Discursive and Visual Urban Frontier in the Branding of the Creative Place

1. Introduction

Arthur Herman coined the term 'declinism' to refer to a kind of deep pessimism that sees the West as horribly flawed, in a state of severe decay, and usually teetering on the edge of catastrophe. Herman finds this radical doubt pervasive both today and in the recent past: "We live in an era in which pessimism has become the norm, rather than the exception." (1997: 2). Herman grounds this theory by citing a considerable range of thinkers who believe that things are getting worse – and getting worse rapidly. For many, the demise of state communism in Central and Eastern Europe tarnished the case for utopia, as attempts to create the perfect society on earth by eliminating poverty, suffering and social injustice, along with establishing the necessary conditions for human flourishing were undertaken in large parts of the world in the twentieth century, and they failed spectacularly.

As the argument goes, the utopian desire to build a better world (and it is often a compelling one) is hubristic and breeds violence, and we are better off without it. In the immediate aftermath of communism's collapse, liberal Western-style democracy on the economic foundation of free-market capitalism was frequently touted as the only viable global solution to the problems that communism has attempted to eradicate as well as the framework for future developments. What Herman did not consider, however, is the persistence of an opposing tradition of devout optimism. The specters of financial meltdown, international terrorism and the global ecological crisis have since combined to render even this anti-utopian utopia untenable.

Nevertheless, the rush to declare utopia dead is premature. Profound instability, rapid change, the bankruptcy of established systems and ideologies — these are, in fact, precisely the conditions under which the concept of utopia has flourished in the past. The grand old idea and myth of progress still has its adherents, and their views are extremely well represented in the marketplace of signs: utopia remains a tenacious and diverse concept in the human imagination. Reports of the death of utopia have been greatly exaggerated (Saage 1990; Fest 1991; Jacoby 1999; Gray 2007). If anything, the death of utopia is a conceptual pawn in the broader game of redefining the political and intellectual traditions of the left and right in the post-communist world, as the case of Berlin demonstrates.

2. Berlin's Post-Unification

After the fall of the Berlin Wall and the reunification of Berlin in 1989, a decade of intense and rapid urban development took place across the city. Berlin can be considered an atypical and extreme case: it is atypical because of its unique and peculiar history as a divided city in a divided country, and extreme because of the intensity of the urban restructuring processes which unfolded over a short period of time following reunification. The acceleration of history represented by the fall of the Berlin Wall and the sudden absorption of East Germany into the capitalist democracy of the Federal Republic of Germany brutally confronted the city with the economic, social, and political challenges faced by many other Western cities over the past several decades. At the same time, Berlin is a fascinating laboratory of urban change that illustrates several (partially interrelated) transitional processes: the transition to a unified city after a history of conflict and division; the transition to a capital city in a nation redefining its national identity; the transition from a socialist to a capitalist city; and the transition from an industrial to a post-industrial or post-Fordist metropolis.

It is precisely because of the peculiar situation of the city that a flurry of practices of place marketing and urban imaging suddenly appeared, with a kind of visibility and intensity rarely witnessed in other (European) cities. In the mid-1990s, visitors to the city's central areas were greeted with an endless landscape of cranes and construction sites. Equally striking was the highly visible presence of images and texts surrounding the construction sites. Public-private partnerships were set up specifically to market the 'new Berlin' to different target groups, including potential investors, tourists and Berliners themselves. Throughout the 1990s and early 2000s, the production of the new buildings in Berlin's reunified urban environment was also accompanied by the construction of a particular image and meaning. This was part of the political responses to the enormous challenges unleashed by the fall of the Berlin Wall and subsequent reunification of the city, ranging from responses to the loss of the political status of exception, the retrieved status as capital city, intense economic restructuring and deep social and demographic transformation. Place marketing refers to "the various ways in which public and private agencies — local authorities and local entrepreneurs, often working collaboratively — strive to 'sell' the image of a particular geographically-defined place, usually a town or a city, so as to make it attractive to economic enterprises, to tourists and even to inhabitants of that place" (Philo and Kearns 1993: 3).

More recently, the term 'place branding' has become increasingly popular, referring to a process of "forging of associations" between a place and some desirable qualities that resonate with particular target audiences (Kavaratzis and Ashworth 2005). This "forging of associations" can be achieved through physical interventions in the city's landscape through forms of communication that reference particular aspects of local identity, history, and culture. This is, in essence, "a highly selective process that imposes single-stranded images onto urban diversity and reduces place identity to a constricted and easily packaged 'urban product'" (Broudehoux 2004: 26). The two terms of 'place marketing' and 'place branding' are often used interchangeably in both professional and academic literature. In any case, the production and diffusion of images is an absolutely central component because "the process of constructing visually based narratives about the potential of places [...] a process of brokering the best metaphor, in ways that will shift or consolidate public sensibilities and invent the possibility for new kinds of place attachments" (Bass Warner and Vale 2001: xv). The image of the city can thus be defined, in a simple way, as having two components: "the physical image of the city — the actual city itself, as it is produced, lived and experienced by people on an everyday basis and represented in a series of visual symbols, physical places, and social characteristics — as well as the rhetorical image of the city — the 'idea' or conceptual image of the city as it is imagined and represented in collective consciousness." (Broudehoux 2004: 26)

In this sense, 2001 marked a turning point in Berlin's post-unification history. In the wake of the exposure of a large-scale financial scandal involving Berlin's public authorities, the Grand Coalition, which had ruled the city for a decade, was replaced with the new 'Red-Red' coalition between the Social Democrats (SPD, *Sozialdemokratische Partei Deutschlands*) and the Left party (PDS, from 2007 renamed *Die Linke*). During its decades as a walled-in island surrounded by GDR territory, West Berlin was heavily subsidized by the Federal Republic and almost as socialist as the eastern part of the city, which was pampered by the socialist state, but the money dried up after unification. The legacy of this was a huge bureaucracy and an ingrained welfare mentality.

Post-reunification Berlin experienced a dramatic economic downturn. East Berlin's inefficient industry — a legacy of its former communist times — crumbled, and West Berlin saw an exodus of companies that no longer received the same tax breaks granted while the Wall had isolated the city. Important firms that moved out after 1945, such as Siemens and Deutsche Bank, saw no reason to return. The state of the city's finances was disastrous and the subsequent restructuring of the local state had a long-term impact on the public services and the welfare of Berliners. More recently, Berlin – not just as a city, but also an autonomous region within the

German federal system – remains heavily supported by the southern regions, which are constitutionally obligated to subsidize the poorer federal states in the north. The consequences of this financial situation are visible as soon as one moves beyond the smart government district around the Brandenburg Gate, or the posh neighborhoods located near the *Kurfürstendamm*. Signs of poverty are everywhere. One third of children in Berlin are poor. Yet in spite of the lack of financial resources, and the constraints on public spending that the new government inherited, the Red-Red coalition continued, and in some cases even intensified, the place marketing activities that had been developed by its predecessors.

3. Sex in the City: Utopian Desire

After the failure of the marketing visions of Berlin as an Olympic city and global service metropolis (Wolf 2000), city leaders struggled to find an alternative vision, but did not give up their search for a master narrative. The new mayor of Berlin, Klaus Wowereit, began to take a very proactive role in the external promotion of the city and worked to make Berlin's imagery support its economic promotion. Just before Klaus Wowereit became mayor in 2001, he coined one of his best-known catchphrases "I am gay and that's just fine" ("Ich bin schwul, und das ist gut so") delivering this announcement to the party convention (SPD) that nominated him for the capital city's top job. With his coming out, Wowereit wanted to beat the tabloids to it and prevent them from publishing wild, sensational and fabricated stories about his private life. This move ultimately strengthened his campaign. His election as mayor made Berlin one of three major European cities with an openly gay mayor, along with Paris, whose mayor at the time was Bertrand Delanoë, and Hamburg, whose mayor was Ole von Beust, who also took office in 2001. However, von Beust resigned in 2010 and Delanoë left office in 2014, making Wowereit the only gay mayor of a major European and German city. As the largest city in Germany and a German federal state in its own right, being the mayor of Berlin also made Wowereit a state premier. His openness on the subject helped foster an atmosphere of tolerance in the German mainstream. It also helped Wowereit to become the face of the tolerant, easygoing, and cool Berlin of the twenty-first century. "Wowi", as he was widely known, was a fashionable, charming mayor with a flamboyant nature and a penchant for rubbing shoulders with artists and celebrities. During his political mandate, he earned a reputation as a hard-working, hard-partying mayor and saw partying and networking as an official part of his job, showing a very human face of socialism.

His big task was to help Berlin avoid the 1990s fate of Washington, DC, which became a bankrupt city with a rich political ghetto. In spite of the nominally left-wing political leaning of his new government, and despite a rhetorical commitment by the governing coalition to maintain social-democratic principles, including preserving high levels of social security, solidarity and welfare, cuts in public expenditure and administrative reforms were prioritized to create a 'service-oriented', 'competitive' and 'business-friendly' city. These cuts were actually a continuation of the previous government's policies and the Red-Red coalition did not significantly shift away from urban entrepreneurial strategies.

With unemployment soaring to around 20 percent, Berlin had to reinvent itself. It was Klaus Wowereit who pushed "the visual and discursive urban frontier" (Smith 1996) in the official representation of the city for marketing and branding purposes in order to accrue distinct "collective symbolic capital" (Harvey 2001). In his discussion of the role of culture in urban entrepreneurialism, Harvey stresses the constantly changing, never-ending nature of the search to maintain a monopolistic edge over urban competitors via the appropriation of local culture(s). In order to support Berlin's transformation into the envisioned European metropolis, local policy-makers had to break away from the negative images associated with the city's turbulent historical past, and reinvent and spread a new image of the city to three main target groups: (1) investors, visitors, and potential tourists; (2) Germans throughout the Federal Republic; and (3) Berliners themselves. For this reason, the mayor turned to a concept of utopia to keep Berlin unique and particular enough "to maintain a monopolistic edge in an otherwise commodified and often fiercely competitive economy" (Harvey 2001: 396–397). In a country undergoing a process of transition between two political systems, and in a city haunted by the specters of its troubled past searching for economic competitiveness on the global stage, Wowereit dramatized emotional attachments to the daily details of a purposeful way of life for the future imagined city and replaced Berlin's financial squalor with the beauty and joy of sex, sketching the utopian scenario of a city that "is poor but sexy" ["Berlin ist arm, aber sexy"]. Under this slogan, lack or deficiency is not necessarily negative or deserving of pity. This is much in contrast to the general message from socialists, who work not only to critique the structural causes of poverty but to build a society that one day will abolish exploitation and poverty. Accordingly, nobody would expect a message from the unofficial crown prince of the Social Democrats that poverty is appealing, attractive, and even sexy.

4. Dreamscape: Selling Propaganda

In *The Principle of Hope*, Ernst Bloch argues that utopian desire — which, as the philosopher demonstrates, is ubiquitous and enduring throughout human history and culture — requires the guiding light provided by socialism, which he calls "the practice of concrete utopia" and "the last chapter of the history of the world" (Bloch 1995: I, 17; 174). He believes in the infallibility of the Marxist analysis of history and society, and its inexorable trajectory towards revolution. His Marxist framework leads him to distrust the emphasis on the processual and open-ended dimension of utopian striving, one which his own work suggests. On the one hand, Bloch seems to acknowledge the fallibility of ideals and the need to mediate between the imagined goal of social change and the flesh-and-blood social reality that approximates this goal. He warns against the "reification of the goal dream" (I, 186), and against the denigration of reality because of its failure to measure up to ideals or dreams: "The dream as such does not realize itself, that is a minus, but flesh and bones are added to it, that is a compensating plus" (I, 187). On the other hand, Bloch also cautions against the opposite problem: the denigration or disregarding of an ideal or dream because of its distance from reality. Bloch may have insisted that Marxist revolutionary theory was the non-negotiable core of his utopian thinking. Yet his omnivorous account of the ubiquity of the utopian moment itself, with its emphasis on the subjective factor of hope, seems to resist being subsumed into a unitary perspective. So much of *The Principle of Hope* works against Bloch's own insistence that could be a definitive path to, and an irrefutable definition of, the utopian "Heimat" or homeland "in which no one has been".

Key Blochian terms such as anticipation, expectancy, the not-yet-conscious, front, and horizon all demonstrate the importance of the processual and temporal dimension of the utopian: it is a future-oriented striving that seeks to attain the unknown by unfolding the possibilities latent within the known. Even where Bloch admits the possibility of a definitive arrival at a utopian goal, a tension remains between the goal as visualized and the goal as achieved. The aporia of fulfilment is about the dialectical relationship between the real and the imagined, about the risks involved in positing a goal: the goal petrifies, becomes static, and loses its connection to the reality towards which it strives. When this happens, the dialectical mediation between them breaks down. In a post-totalitarian age, utopian desire becomes a beginning that is denied an ending, a movement that is cheated of and yet continues to suggest its destination (Bauman 1991: 244; Frank 1979). The intransitivity that defines a movement without a goal captures a key feature of the utopian in its postmodern incarnation. The 'intransitive' refuses a definitive formulation of aims or objects, emphasizing instead the need to continuously revise these in an endless process of approximation. The intransitivity of postmodern utopianism stems from its insistence that while a definitive account of the destination is admitted to be impossible, the necessity of continuing the journey cannot be denied. The object of utopian striving may recede from view or resist formulation, but this does not negate the striving itself.

A central problem identified by critics of utopianism becomes apparent in the contrast between modern transitivity (i.e. the static, potentially totalitarian vision of the end) and post-modern intransitivity: the problem of the relationship of means to ends. From Max Horkheimer and Theodor W. Adorno (2002) to Ivan Illich (1973), several thinkers have identified the emancipation of means from ends as a key problem of the modern condition. The enlightenment that is to liberate man from myth itself takes on the oppressive qualities of myth; the tools man creates to replace slaves begin to enslave him. In these examples, the means become oppressive when they outgrow or become dissociated from the ends that they were originally developed to attain. The aporetic relationship of means to an end is formulated quite differently by antiutopian thinkers such as John Gray (2007). In the view of antiutopian thinkers, it is the end that is a source of tyranny when it is used to justify any means. From this perspective, the danger of utopia lies in the requirement that the present be sacrificed to the future.

In this sense, Wowereit's pragmatist approach treated Berlin as an imagined place where specific ways of life reconcile current problems while suspending commitments to many current constraints. His second famous, oft-cited catchphrase was "the negation of a negation", "the absence of radical evil" (Kateb 1973: 242, 240): the vision of a new land of Cockaigne where money does not matter, but joy is secured, a city that is cool and cheap, drunk and druggie, slightly anarchic and not pompous, penniless for the time being but open, young, and full of ideas with a radiant future — very unique and different, at least, from the rich but saturated and rather boring cities like Munich or Hamburg. It was the semiotician A. J. Greimas who made abundantly clear how the presence of any value in language must invariably produce its negation (Greimas and Courtés 1979: 308–311; Greimas 1983). In this sense, poverty becomes, in Wowereit's words, an important resource, a sign of distinction and of the arrival of a new "libidinal economy", to misappropriate a phrase used by Jean-François Lyotard (1993).

Selling utopia is one of the cultural roles of contemporary propaganda. Although this is hardly the obvious purpose of civic advocacy, it is indeed one of its inadvertent effects. The nature of the game requires that civic advertisements convey a yearning for a better world. This is an effective way to reach out to the public. Rarely can the sponsors promise the kind of tangible rewards that come with purchasing a brand and its image. They seek to educate people, encourage them to open their wallets, and often to alter their behavior, each a far more difficult task than simply shaping their choices as consumers. Accordingly, sponsors strive to connect their messages to the broader utopian impulse or propensity in the public, to imply, and sometimes to show, how they will remedy a wrong, avoid some evil, confirm a value, and so on. Borrowing here from the language of Saint-Simon via Ricoeur, they try "to impassionate society" in order to "move and motivate it" (Ricoeur 1986: 296). Embedded within civic advocacy is that "inner dialectic of utopia, its rational and emotional sides" (Ricoeur 1986: 287). Appeals draw upon an eclectic variety of desires for abundance, community, power, freedom, and peace.

5. Berlin as a Creative Place

Given the poor state of the city's public finances, Wowereit's utopian scenario operated on three levels: (1) as fantasy and escape; (2) as an alternative, a challenge, a reaction to the existent; and (3) as innovation, exploring the possible, a source of extreme novelty. Utopia is simultaneously an ideal, desire, and critique. That ideal is concrete; or, rather, it is expressed in ways that give it specificity, at best making the ideal appear both unique and bold, an aesthetic triumph. It bears a signature: "Utopias are assumed by their authors", claimed Paul Ricoeur, "whereas ideologies are denied by theirs" (Ricoeur 1986: 2). It was the French philosopher who argued the significance of contemporary life, utopia as fancy, as an alternate to the present power, as the exploration of the possible. Berlin's carnival of sex referred to an erotic urban paradise ruled by the priority of appetite, a place fascinated with sin and excess, committed to play and indulgence. It presented an alternative realm of existence where people might escape the trials and tribulations of their everyday lives. The new urban regime of stimulation would soon prove to be a much more pleasing mode of governance than the earlier brand of repression, largely due to the fact that it allowed room for play and pleasure, even a modicum of rebellion. Perhaps more striking, Wowereit's Eros project served, like so much of pop culture, to reinvigorate a world rendered prosaic and dull by the rise of logic and industry.

Urban image construction is "an objective and productive social force, with real material effects, playing an integral role in shaping modern forms of production, consumption, and collective 'dreamscape'" (Greenberg 2008: 20). "A utopia is not only a dream but a dream that wants to be realized. It directs itself toward reality; it shatters reality" (Ricoeur 1986: 289). Adopted by Berlin's Red-Red coalition as master narrative of the 'creative city' discourse of urban policies and place marketing, utopia invited the audience to journey to a unique ideal place, leaving behind the current troubling details of the present. Berlin as an imagined space for settlement and investment casts the difficult situation in the present as form and type of accomplishment that can replace familiar problems with new ways of life, and distinguish itself on the global economic stage. In this sense, a special emphasis is put on providing active support (and marketing) to the creative industries or the creative economy (of which cultural production is an important part). The theme of creativity has been adopted as a focus of local economic development policy and as a marketing slogan by many urban policy-makers across the world on the basis of the analysis and police recommendations made by Charles Landry (2000) and Richard Florida (2002).

Initially developed with regard to the cities in the United States, Florida's main argument is that economic growth and innovation are now driven by the "creative class", which he divides into the "supercreative core" of professionals "whose economic function is to create new ideas, new technology and/or creative content" (ibid., 8) in the fields of the natural sciences, engineering, computer programming, education and research, the arts, design and media; and the "creative professionals", workers in the knowledge-based industries such as business, finance, law, healthcare. In the new "creative economy", he argues, cities, not nations, compete for highly mobile 'talent', meaning that workers can choose the places where they would like to live. The characteristics of a city sought after by the members of the "creative class" are, among others, a vibrant cultural life, and a tolerant and unique atmosphere. The policy implications of this thesis are that urban policies should assist in creating the conditions for attracting the creative classes by supporting the formation of a "creative milieu" "that contains the necessary preconditions in terms of 'hard' and 'soft' infrastructure to generate a flow of ideas and interventions" (Landry 2000: 133). While Florida's thesis, assumptions, definitions and arguments have been widely criticized by academic studies (Montgomery 2005; Peck 2005; Markusen 2006; Krätke 2010), his ideas have had a spectacular impact on urban policymakers. Berlin's mayor was no exception, and he began to work to promote Berlin as appealing and "creative" on the basis of the strong growth of the cultural industries in the city.

According to the first comprehensive study of these industries carried out by the Senate Department for Economics in 2005 (which encompassed the sectors of publishing, print media, film and TV production, fashion, design, software and games development, telecommunications, music, advertising, architecture and exhibition arts), this sector has been the fastest growing sector in the city's economy since the late 1990s, at rates higher than in other German *Länder* (SenWi 2005). An updated report on Berlin's cultural economy published in 2008 estimated that by 2006, the sector accounted for 10 percent of the workforce and 21 percent of the city's GDP (SenWi 2008). Looking more broadly at the growth of knowledge-intensive industries in Berlin, researchers have shown that Berlin has developed a comparatively strong position at the European and even global level in several subsectors such as media, software and the life sciences (Krätke 2002, 2003, 2004a, 2004b). Krätke argues that Berlin has become an "alpha world media city", "a center for cultural production and the media industry with a world-wide significance and impact" (Krätke 2003: 618; Krätke and Taylor 2004) on par with Los Angeles, New York, Paris, London, Munich and Amsterdam.

Given this, there is a noteworthy discrepancy between Berlin's rather low position in the global system of strategic economic centers and its leading position in media production of cities around the world. Indeed, Berlin has become a hotbed of cultural innovation not despite of, but perhaps because of its weak economic performance in the conventional sense (Bader and Scharenberg 2010). The growth in cultural industries in Berlin was largely unplanned, facilitated by the availability of affordable working and living spaces, by a tolerant and liberal culture inherited from the 1970s and 1980s, and supplied with preexisting concentrations of cultural producers, artists and networks of alternative culture (e.g. the techno music scene).

As the Berlin Senate became increasingly aware of the role and potential of the cultural industries in the local economy, it began to develop various programs and policy initiatives to support new business startups, and the city produced some globally recognized tech startup successes, including the music sharing service SoundCloud and games company Wooga. It also promoted interfirm networking, and encouraged 'creative clustering' in underutilized urban spaces (Ebert and Kunzmann 2007). The targeted sectors included Information and Communication Technologies, film, TV and radio, print and digital media, music, fashion, design, art and architecture. Creative industries have been valued as an economic sector in their own right, as a location factor for (other) knowledge-oriented companies, young creative entrepreneurs and their workers, and as an attraction for (young) urban tourists.

6. Berlin as the New Neverland

Apart from hard location factors (such as the availability of affordable spaces), the city's lively club and music scene, gay culture, hedonistic nightlife, multiculturalism and tolerance were increasingly integrated into the mainstream marketing discourse as unique selling points for Berlin. Mayor Wowereit (2006) quoted Florida's trio of urban virtues, ("technology, talent, and tolerance") in his inaugural speech to the House of Representatives at the beginning of his second term. In the case of Berlin, this process of symbolic appropriation has been fueled by the media, as illustrated by *Time Magazine*'s November 16, 2009 headline "Hip Berlin,

Europe's Capital of Cool". The combination of a changing policy focus and of changing cultural and consumption practices explains why new sites and spaces have become integrated into the formal representation of the city to the outside world, accompanied by a narrative of 'creativity', 'diversity', 'tolerance' and 'hipness'. Berlin's poverty, marginality, and purported misery, which were previously seen as weaknesses, were reimagined in different and interesting ways and promoted as strengths to specific target audiences. If the elitist character of the subculture (e.g. through secretive happenings with restricted access) "fits the self-stylization of the new urban middle-classes" (Bader and Scharenberg 2010: 84–85), then the possession of "subcultural capital" signals status in the form of "hipness" (Thornton 1997), which is based on a constant renegotiation and extension of the boundaries of legitimate culture to include new, previously illegitimate artistic and cultural forms. Gradually, symbols of youth and alternative cultures such as the Love Parade, Christopher Street Day or the Carnival of Cultures were integrated into tourism and place marketing. These urban festivals initially emerged from relatively marginalized groups (techno fans, gays and lesbians, and migrant organizations), but were later officially marketed as part of the desire to present Berlin as a young, tolerant and cosmopolitan city (Kalandides and Lange 2007: 128) The urban voids and wastelands – vacant or abandoned lots that suffered from war damage, Cold War era division, poor planning decisions, demolitions by successive political regimes or deindustrialization — were previously left out of the promotional imagery as signs of desolation and traces of unwanted past. As part of the new Berlin imaginary, these spaces were domesticated by virtue of their utopian innovative temporary potential, and as new playgrounds for artistic production, consumption, creativity, entertainment and leisure for creative pioneers providing a unique selling point for Berlin. The city alongside the river Spree and the canals were turned into a long beach complete with sand, deck chairs, exotic decorations and music or an outdoor swimming pool (*Badeschiff*), and the abandoned lots were occupied by new bars and clubs, which became particularly popular images of a new German Neverland. Berlin's rather ordinary, socially mixed neighborhoods (such as Kreuzberg), with their comparatively high concentrations of unemployment and poverty, high proportion of foreign-born or minority background residents (e.g. Turkish, Arabic, Central, Eastern and South-Eastern European) and their authentic as well as alternative and counter-cultural lives, have increasingly been portrayed in marketing campaigns and publications as tourist attractions or potential settings for young creative entrepreneurs. The promotion of 'cosmopolitanism' through particular spectacles of identity, such as the Carnival of Cultures or the Turkish Parade, employs markers of ethnicity, culture, gender, youthfulness, hip and coolness to celebrate cultural particularisms within an imaginary of diversity.

The prioritization of urban tourism and the growing demand from tourists for encounters with remnants from the Nationalist-Socialist and Cold War eras of Berlin has led city marketers to integrate into the promotional imagery historical sites and traces (authentic or purposefully recreated) that were largely left out or even concealed in the marketing discourse of the 1990s. A new 'memory district' has emerged in the center of Berlin, which includes the Jewish Museum (opened 2001), the Memorial for the Murdered Jews of Europe (inaugurated in May 2005), and the topography of terror, a site where the unearthed remains of the Gestapo headquarters have been on public display since 1987, and since 2010, a new documentation center provides information on the National-Socialist past. Although the 'memory district' was not planned or marketed as such, it gradually evolved as a coherent tourist concept, as stressed by Till (2005: 200). It is a perfectly magical stage of remembrance. The remains of the Wall also slowly transformed from "sites of dispute" to "sites of memory" (Dolff-Bonekämper 2002). In this way, the dark periods of Berlin's past were reframed within the utopian scenario of a sexy, creative city as a very open interaction with its own history that creates hyper visible space within the city for new ways of thinking.

7. Conclusions

In this analysis of political and cultural forces behind the slogan heralding a "poor but sexy" Berlin, a surprisingly grim side to paradise is revealed. There is a fundamental contradiction at the heart of this process of pushing the discursive and visual urban frontier in Wowereit's branding strategy: "using a city's subculture may enhance the city's symbolic value, but simultaneously undermines the everyday conditions necessary to sustain the creative process itself" (Bader and Scharenberg 2010: 80); reducing interim and small-scale users to a marketing tool for real estate in the city is detrimental to a proper long-term creative city policy. This, in turn, generates resistance on the part of cultural producers and users who are affected by the process, and leads to localized conflicts around the spaces promoted as 'creative' in public policies and in official marketing discourse. Kreuzberg, before the fall of the Wall, was already a hub for Berlin's alternative, multicultural and bohemian scenes, and as such has attracted visitors not only from Berlin, but also from the rest of Germany and abroad since the 1970s (e.g. Iggy Pop and David Bowie). After the fall of the Wall, the district underwent an intensive process of urban renewal, social transformation and, of course, gentrification. There was and is a discrepancy between the mayor's and senate's promotional rhetoric of ethnic diversity and sexy cosmopolitanism on the one hand, (a central element of Berlin's marketing narrative in the 2000s), and its actual management of ethnic-cultural diversity and inequalities on the other (Kosnick 2009). Finally, the increasingly popular and profitable marketing of memory has normalized the darkest periods of Berlin's history and passed over in silence the negative consequences of German unification on (East) German society and the failures of the process.

References

Adorno, Th. W. and M. Horkheimer (2002). *Dialectic of Enlightenment: Philosophical Fragments*. Stanford: Stanford University Press.

Bader, L. and A. Scharenberg (2010). "The sound of Berlin: subculture and the global music industry". *International Journal of Urban and Regional Research* 34(1): 76–91.

Bass Warner Jr., S. and L. J. Vale, (2001). "Introduction: cities, media, and imaging". In: Bass Warner Jr., S. and L. J. Vale (eds) *Imaging the City: Continuing Struggles and New Directions*. New Brunswick, NJ: Rutgers University, Center for Urban Policy Research, xiii–xxiii.

Bauman, Z. (1991). *Modernity and Ambivalence*. Ithaca: Cornell University Press.

Bloch, E. (1995). *The Principle of Hope*, 3 vols. Cambridge, MA: MIT Press.

Broudehoux, A. M. (2004). *The Making and Selling of Post-Mao Beijing*. London: Routledge.

Dolff-Bonekämper, G. (2002). *The Berlin Wall. An archeological site in progress*. In: Johnson, W. G. and C. M. Beck (eds) *The Archeology of 20th Century Conflict*. London: Routledge: 236–248.

Ebert, R. and K. Kunzmann (2007). "Kulturwissenschaft, kreative Räume und Stadtentwicklung in Berlin". *disP* 171: 64–79.

Fest, J. (1991). *Der zerstörte Traum: Vom Ende des utopischen Zeitalters*. Berlin: Siedler.

Florida, R. (2002). *The Rise of the Creative Class*. New York: Basic Books.

Frank, M. (1979). *Die unendliche Fahrt: ein Motiv und sein Text*. Frankfurt a. Main: Suhrkamp.

Gray, J. (2007). *Black Mass: Apocalyptic Religion and the Death of Utopia*. London: Penguin / Allen Lane.

Greenberg, M. (2008). *Branding New York: How a City in Crisis Was Sold to the World*. London: Routledge.

Greimas, A. J. (1983). *Structural Semantics: An Attempt at a Method*. Lincoln: University of Nebraska Press.

Greimas, A. Julien and J. Courtés (1979). *Semiotics and Language: An Analytical Dictionary*. Bloomington: Indiana University Press.

Harvey, D. (2001). *The art of rent: globalization and the commodification of culture, in D. Harvey, Spaces of Capita*l. London: Routledge, 394–411.

Herman, A. (1997). *The Idea of Decline in Western History*. New York: Free Press.

Illich, I. (1973). *Tools for Conviviality*. London: Calder & Boyars.

Jacoby, R. (1999). *The End of Utopia: Politics and Culture in an Age of Apathy*. New York: Basic Books.

Kalandides, A. and B. Lange (2007). "Creativity as a synecdoche of the city – Marketing the creative Berlin". In: Wan, W., E. Yueng and T. Yueng (eds) *Hong Kong Institute of Planners & Urban Planning Society of China Conference 'When creative industries crossover with cities'*. Hong Kong, 2–3 April. Available at <http://www.hkip.org.hk/ci/paper/Ares%20Kalandides.pdf>.

Kateb, G. (1973). "Utopia and the Good Life". In: Manuel, F. (ed) *Utopias and Utopian Thought*. London: Souvenir Press: 239–259.

Kavaratzis, M. and G. J. Ashworth (2005). "City branding: an effective assertion of identity or a transitory marketing trick?" *Tijdschrift voor Economische en Sociale Geografie* 96(5): 506–514.

Kosnick, K. (2009). "Cosmopolitan capital or multicultural community? Reflections on the production and management of differential mobilities in Germany's capital city". In: Nowicka, M. and Rovisco, M. (eds) *Cosmopolitanism in Practice*. Aldershot: Ashgate: 161–180.

Krätke, S. (2002). *Medienstadt. Urbane Cluster and globale Zentren der Kulturproduktion*. Opladen: Leske & Budrich.

Krätke, S. (2003). "Global media cities in a worldwide urban network". European Planning Studies, 11(6): 605–628.

Krätke, S. (2004a). "Economic restructuring and the making of financial crisis: Berlin's socio-economic development path 1989 to 2004". *disP* 156, 58–63.

Krätke, S. (2004b). "City of talents? Berlin's regional economy, socio-spatial fabric and 'worst practice' urban governance". *International Journal of Urban and Regional Research* 28(3), 511–583.

Krätke, S. (2010). "Creative cities' and the rise of the dealer class. A critique of R. Florida's approach to urban theory". *International Journal of Urban and Regional Research* 34(4): 835–853.

Krätke, S. and P. J. Taylor (2004). "A world geographies of global media cities". *European Planning Studies* 12(4): 459–477.

Landry, C. (2000). *The Creative City: A Toolkit for Urban Innovators*. London: Earthscan / Kogan Page.

Lyotard, J.-F. (1993). *Libidinal Economy*. Bloomington and Indianapolis: Indiana University Press.

Markusen, A. (2006). "Urban development and the politics of a creative class: evidence from a study of artists". *Environment and Planning* A 38(10), 1921–1940.

Montgomery, J. (2005). "'Beware 'the creative class'. Creativity and wealth creation revisited". *Local Economy* 20(4): 337–343.

Peck, J. (2005). "Struggling the creative class". *International Journal of Urban and Regional Research* 29(4): 740–770.

Philo, Ch. and G. Kearns (1993). "Culture, history, capital: a critical introduction to the selling of places". In Kearns, G. and C. Philo (eds) *Selling Places: the City as Cultural Capital, Past and Present.* Oxford: Pergamon Press: 1–32.

Ricoeur, P. (1986). *Lectures on Ideology and Utopia*. New York: Columbia University Press.

Saage, R. (1990). *Das Ende der politischen Utopie?* Frankfurt a.M.: Suhrkamp.

SenWi (2005). *Kulturwirtschaftsbericht*. Berlin: SenWi.

SenWi (2008). *Kulturwirtschaft in Berlin. Entwicklungen und Potenziale. 2. Kulturwirtschaftsbericht*. Berlin / SenWi: Senatskanzlei, SenStadt.

Smith, N. (1996). *The New Urban Frontier: Gentrification and the Revanchist City*. London: Routledge.

Thornton, S. (1997). "The social logic of subcultural capital". In: Gelder, K. and S. Thornton (eds) *The Subcultures Reader*. London: Routledge: 184–192.

Till, K. E. (2005). *The New Berlin: Memory, Politics, Place*. Minneapolis: University of Minnesota Press.

Wolf, H. (2000). "Zur Kritik der Metropolenpolitik des Berliner Senats". In: Scharenberg, A. (ed.) *Berlin: Global City oder Konkursmasse? Eine Zwischenbilanz Zehn Jahre nach dem Mauerfall*. Berlin: Dietz: 146–154.

Wowereit, K. (2006). *Regierungserklärung des regierenden Bürgermeisters Klaus Wowereit*, Abgeordnetenhaus von Berlin, 22 February. Berlin: Senatskanzlei.

2. Economics and Communication

Paola De Vecchi Galbiati

Self-adaptive Organisms: The Evolution of Organizational Models and Systems

1. Self-adaptive Organisms

> A Self-Adaptive Organism is a system able to reconfigure its own structure and change its own behaviour during the execution of its adaptation to environmental changes. (Zhu 2015)

The function of a self-adaptive organism can be described in a recursive way: this entity updates its results, means and processes by thinking and acting, recycling knowledge to solve problems both inside and outside of itself.

Self-adaptation is an observable phenomenon: nature provides numerous examples of self-adaptive organisms. These models are used in biology, physics and chemistry as metaphors to create reference systems and metrics that allow us to improve and enhance organizational solutions in social and economic contexts.

2. Self-Adaptation in biology

When observing adaptive organisms, biology focuses on processes at the sub-organism level. Adaptations of organisms to environmental stresses are studied and developed using the latest molecular and physiological tools. Regulatory mechanisms are understood, including genetic constraints, physiological plasticity and the evolutionary history of the responses of organisms.

In biology, the sheer multitude of self-adaptive solutions implemented by natural organisms is overwhelming: plants, animals and micro-organisms convey crucial information for understanding the effects of an environmental change on the organisms, as well as the effects of organisms on the environment change.

This information together provides insight into how different communities (plants, animals, microorganisms) are interrelated and interdependent within an ecosystem, and how they determine the resilience of communities to environmental stress (Folke et al. 2007).

3. The Brain is a Self-adaptive Organism

Brain plasticity refers to the innate ability of the brain to adapt by forming new neural connections and strengthening synapses between existing neurons during the course of our lives. This occurs in response to experiences, acquiring skills and learning. This also includes the exchanges of information between people and their cultural environment (Chiel and Beer 1997).

The ability to adapt allows us to assimilate new information and to create new memories. The most interesting aspect of the brain's plasticity is its ability to "put knowledge into operation", using everyday situations as catalysts for self-adaptation (Scott Kelso 1997).

4. Self-adaptation in Design and Engineering

Solutions in automation, robotics, bioengineering, artificial intelligence and information technologies replicate self-learning and self-organization – abilities that can be observed in natural organisms.

Designers and engineers imitate the behavior of autonomous organisms in designing and developing self-adaptive products and services. These systems are developing and autonomous structures that are able to store behaviors and situations, to exchange information with the environment and with other systems, and to select data and use it in appropriate contexts (Schönera et al. 1999).

Industrial automation has evolved so far that equipment and machinery are able to perform self-diagnosis and self-repair. These systems are equipped with artificial intelligence that can store and play behaviors due to experiences stored and shared continuously during their operations.

If we think about the technology services with which we interface daily, we see that they collect our behavior incrementally. Through data mining and user interfaces, they self-adapt themselves to us; digital media suggests products and services that match our tastes or our interests.

5. Self-adaptation in Sociology

Self-adaptive models are also being used in business organization and management. During these last five years, new organizational models for managing teams and projects have been designed and implemented (Zamaraeva and Pankratov 2013).

By applying biological models to changing organizations, we have seen that self-determination, self-regulation and self-control of individuals can increase rapidly in certain conditions. And by thus achieving a high level of autonomy, each person becomes a catalyst for self-adaptation skills in their entire group.

In my research, I have considered three different approaches to evaluate self-adaptation in sociology:

- Self-adaptation of groups when the environment changes.
- Self-adaptation of groups when objects and services change.
- Self-adaptation of groups when rules and roles change.

Considering the first point, there are many ways in which cultural changes have been induced by adaptation to environmental changes. Reduction of natural resources such as water and fertile land generates migration to more habitable places, and at the same time, self-adaptation in consumption for those who remain, for those who emigrate, for those who host. The occurrence of extraordinary natural events (e.g. earthquakes, floods) has produced devastation and simultaneously has refined prevention mechanisms and emergency services. And many people are working on the adaptation of organizational models to achieve stable conditions. Although the slowness of the response may be alarming, groups are refining their awareness of the impact of organizational structures and behaviors on the natural environment. As self-adaptive organisms, humans act simultaneously on their own structure (meta-cognition) and behavior (social responsibility).

With respect to the second point, I can see self-adaptation at work everywhere, every day. Many products and services have been launched in the last years, including mobile devices, apps, cloud computing and social media. By using these technologies, we are all changing our mindset and approach: we now use technologies to reserve a table at a restaurant, to buy tickets, to read a book or to watch a movie. This self-adaptation occurs due to a change in habits resulting from technological changes. After this first stage, all our self-adaptation mechanisms are activated: we imitate behaviors and structures learned from technologies, adapt solutions (e.g. resource sharing, peer to peer, e-collaboration), and introduce these solutions to social and economic organizations (e.g. open access, creative commons, crowd-funding).

At the third point I try to provide a view of an organization like an organism (group) of organisms (people), interconnected through "some pieces of culture". These aspects of culture are rules and roles that we can already know or we can learn at home, at school, or in our workplaces. All of us have many stories to tell about times, styles, and ways to adapt ourselves to organizational changes. The renewal of an established culture generates conflicts and tensions between people and groups in adapting our structures and behaviors. Use a self-adaptive approach in managing cultural projects entails delegating power and responsibility to individuals (e.g. employees, managers, students and teachers). In this way, individuals' capacity to adapt themselves improves. In turn, this personal self-adaptive improvement accelerates the self-adaptation of an entire organization. This kind of evolution will be demonstrated in the following chapters.

6. An Evolutionary Perspective in Business and Social Organizations

We are self-adaptive organisms: we reconfigure our structure (e.g. mentality, method, approach, style) and we change our behavior (e.g. relationships, actions, reactions) through the continuous exchange of information with the environment (Smith 2004; Adolphs 2009).

If we see ourselves as natural and cultural entities, with some physical and cultural characteristics, we can also consider our organization as a living entity that includes and merges different people and different cultural entities.

We can share information with our natural and cultural environments and this simple event causes a change: first in our mindset, then in our behavior, and finally in our organizational system.

7. Scope and Approach of This Research

For several years I have worked with many individuals in different profit and non-profit environments and collected extensive data about the self-adaptive skills that people already have and can improve when it comes to facing environmental and cultural changes.

As reference samples, I evaluated three different organizations in three different contexts:

- Volunteers: When we put together people with different cultural backgrounds but driven by common interests, the group's effectiveness and efficiency can increase rapidly. We can enlarge this perspective by describing ourselves as cultural elements of more a complex organism. Using an evolutionary approach, we can see that each individual increases their cultural dimensions and collaborates to reinforce the organism's structure during a continuous and incremental process. This organism evolves its common culture through the exchange of information of its components.
- Secondary School (K–12): When we introduce digital and non-digital tools in education, the different talents and attitudes of students can emerge more easily. Disruptive technologies applied to didactics allow teachers expand their modelling diagram and see digital technologies as new cultural dimensions entering into an established cultural organism: the secondary school. This complex system is made up of teachers, parents, roles, rules, tools and educational programs and this organism now includes new cultural dimensions (digital technologies) and simultaneously adapts itself and its cultural components (people, rules, contents).

- Company Merger: When we involve each person in merging two different companies into a new one, sharing and co-creation are 'self-emerging' attitudes in everyone. Considering two companies like two living entities, we point out cultural elements that can facilitate the adaptation of individuals to the new organism. This organism is born of the cross-fertilization of different cultural dimensions of each person. Employees were involved in a collaborative redesign of the company's processes and rules. People with high levels of engagement are able to adapt themselves and change the solutions through daily practices. Important catalysts in this process include self-learning, self-regulation, self-repairing.

I was part of each of these organizations for 3–5 years, and I had the opportunity to follow them closely. This allowed me to measure the evolution of self-adaptation in both individual people and organizations, observing and identifying the elements that can facilitate change management.

By tracking the structure and behaviors of these organizations, we have identified which behavioral and structural conditions can accelerate or delay self-adaptation. And this has allowed us to design, test and refine methods and techniques to enable self-adaptation.

8. Progetto eXtra

Progetto eXtra is a group of volunteers living in a small town in Italy (Lombardy). There are 6,200 residents and 14 percent are foreigners. The mission of this group of people is to enhance the meeting and merging of different cultures. Group members share the same vision and interest: helping Italians and foreigners understand and and exchange their cultures.

In 2005, the first course in the Italian language started, with five teachers and 40 students from 15 countries in attendance. Every year new students join Italian courses, attracted by the word of mouth from their fellow countrymen and by information shared in schools and public offices. In ten years, this organization has involved more than 400 people from 40 different countries and more than 20 Italian volunteers.

The volunteers also provide assistance to support foreigners in understanding Italian laws and help them interface with public authorities (social assistance, health care services, schools and services for children). The local community also makes efforts to involve foreign residents in cultural events sharing traditions, history, food and music.

This group supports local institutions in promoting the self-adaptation of foreigners to the local environment. All members of *Progetto eXtra* are "change agents" who make self-adaptation feasible and sustainable.

In 2012 and 2014, two volunteers collaborated with *Progetto eXtra* in order to understand how this community "runs" and to implement a similar solution. After one year of collaboration with *Progetto eXtra*, they started up a similar program in their towns. In other words, *Progetto eXtra* was imitated: these two volunteers developed and adapted the structure and behaviors of the 'original solution' to their local environment.

Figure 1

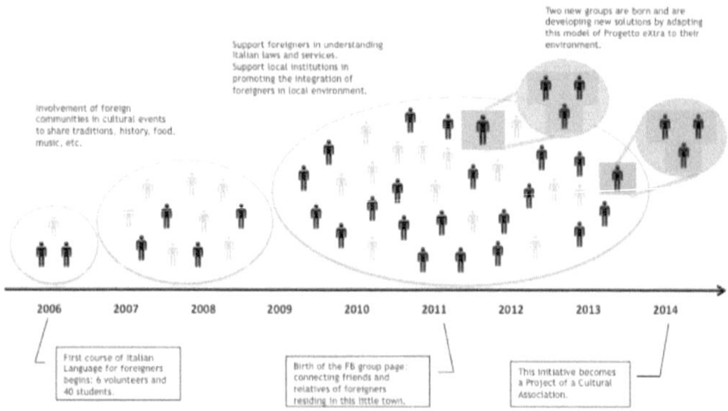

The evolution of this cultural project is summarized in the following key points:

- This group was born in a small community of people strictly linked to the social environment in which they live and work.
- Common interests are clear and explicit; there are no hierarchies or preconceived roles in this organism.
- Collaboration between this autonomous entity and local institutions has been always effective.
- Relationships between the members of this group are based on sharing and collaboration. The contribution of each person is voluntary and related to personal skills and experiences.

9. Generazione APP

Generazione APP is an experiment that tries to create solutions in which the "formal relationship" between the roles of teachers and students is radically transformed to benefit the "natural relationship" between individuals, and between the individual and the group. We want to demonstrate that this approach accelerates learning.

This initiative was born in 2012 at a public school in Italy. The goal was to develop a set of tools and actions that would help extend the offerings for students and the spectrum of evaluation for teachers.

I started this project in a small school near to Milan (Italy) in 2012 with a workshop titled "How Social Networks are changing the Way We Communicate" given to four classes of 12-year-old students.

During the first part of workshop, I shared with the students a set of "key indicators" for using the Internet and social media as "education and learning catalysts".

In the second part of workshop, students explained to teachers how school could be transformed into a more attractive and interesting place. They presented their suggestions using drawings, mock-ups, videos, and slideshows, showing many interesting solutions that could be used to help make schools more engaging and exciting.

The students indicated that self-education models are more engaging, fun and profitable. Students and digital media are the change agents of self-adaptation in the school organism.

Due to the success of this initiative, from 2014 we implemented more than 40 hours of workshops with students and teachers for each discipline (mathematics, history, art, music, Italian, English and French languages, science), with the following common objectives:

- Introducing digital technologies in education and evaluation processes.
- Designing and implementing new instructional formats (Augmented Didactics).
- Reducing the gap between the communication style of adults and pre-adolescents.

The commitment and results shown by all students were very high, and teachers were astonished. We have held four sessions (from 2012 to 2015) of these workshops, involving more than 400 students and 25 teachers.

Figure 2

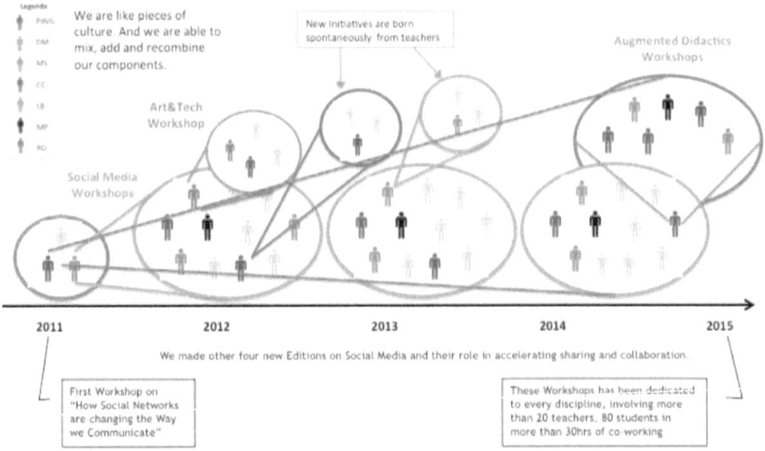

The evolution of this cultural project at a school provided the following findings:

- The idea for this project stems from informal meetings, and not from a 'big plan'.
- The common interest of parents and teachers is clear and explicit: educating the children. And for this reason, sharing and collaboration were and are the basis for the relations.
- Teachers launched digital projects, replicating the model into their sphere of interests and objectives and they self-adapt their own mindsets with regard to this digital revolution in education.
- Our organism (experimental classes and lessons) inherits an organizational code from the environment in which it is inserted.
- To perform these tasks in a public school, there are hierarchies, roles, procedures and standards that have to be respected. And this can result in delays in achieving innovations and changes.

10. Back to School

This last case study summarizes the activities made for merging two small companies into a new company: the first company was an ISP company (delivering solutions for video capturing, streaming and other Internet services), and the second company was a PR agency (press office, event management). The merger of these two organisms generated a new company in integrated communication. The merger took place from 2007 to 2009.
The aims of stakeholders were:

- Expanding the portfolio of services and products offered, which in turn would increase market share in digital PR and web services.
- Accelerating the development of a unique company culture through the mutual exchange of humanistic and technical knowledge.
- Increasing profits by creating streamlined and effective operational processes.

In this working group, I served as the program manager and business coach. This allowed me to introduce 'new pieces of cultural code' that companies did not yet have in their cultural DNA.

More than 50 employees and consultants were involved in a 'learning by doing' educational program.

During these stages, the employees learned methods and techniques for managing projects, developing new products and managing customers. These techniques were then applied to real projects and customers. The employees shared problems and co-created solutions and adapted themselves to a new way of working.

The work of internal reorganization was carried out involving all people to propose solutions and better manage this transformation. In this way, the dedication and enthusiasm of people was very high and as a result, very productive. There were moments of crisis and tension, which were even painful for some.

Nevertheless, feeling included in this process of change made the transformation more tolerable for everybody.

Figure 3

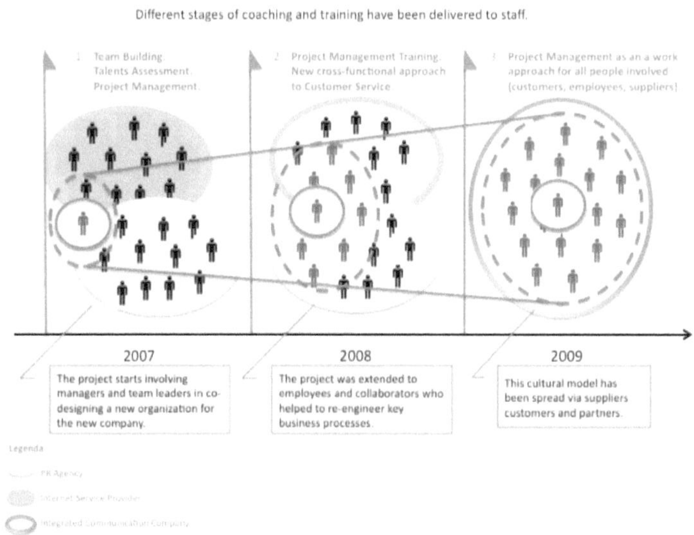

The elements contributing to the success of this project are the following:

- The interests of the two companies were clear and explicit: merging into a single company to increase profits.
- A change agent was included in the group to transmit new skills and methods for operations management.
- Organizations were streamlined, resulting in no overlap of roles and responsibilities.
- New hierarchies and roles were born from the merger and the adaptation of existing ones: the self-adaptation of each person spread out the entire company structure and behavior.
- Involvement of staff in research and development of organizational solutions increased efficiency and effectiveness.
- Concrete results in reducing time to market and project costs encouraged the spread of new practices in accounting and project management in the new company.
- Project management as approach to cooperation encouraged the spread of co-creative solutions: we have shown that project management approach can be considered a change agent for accelerating self-adaptation to internal and external changes.

11. Conclusions

The different groups analyzed had three structural variables in common: group size, the heterogeneity of components and the simplicity of rules.

The relationship between the elements of the groups was based on three "connective" elements: trustworthiness, reputation and reciprocity.

Evolution did not come from a blue print, but emerged from continuous exchange between the elements of the group and the environment.

The hierarchies (do – manage – control – decide) are less reactive than self-organizations that arise spontaneously when people want to find a solution to a problem.

Small changes in the initial conditions brought great benefits to the evolution of solutions.

People co-create solutions, sharing their own pieces of cultural code during the entire transformation process.

Considering organizations as living beings is no longer a simple metaphor, but it is a framework to accelerate and spread our ability to adapt as individuals and as organizations.

The self-adaptation process is a sequence of everyday practices. We refine solutions while we are experiencing them.

Gradually we improve our ability to learn by experiencing our ability to adapt to environmental changes.

In order to be able to adapt ourselves to external changes in a short time and without permanent damage to the structures and functions, we have to work on our skills of understanding and communicating with individuals and the environment.

Expanding our channels of communication simultaneously, we expand our opportunities for data acquisition, processing, learning and adaptation.

These projects and communities show that "co-built" businesses and organizations based on sharing, collaboration and co-creation are not only feasible, but profitable.

References

Adolphs, R. (2009). "The Social Brain: Neural Basis of Social Knowledge". *Annual Review of Psychology* (<http://www.ncbi.nlm.nih.gov/pmc/articles/PMC2588649/>).

De Vecchi Galbiati, P. (2012). "Reinvent and Reorganize the Work in Time of Crisis. A Memetics Perspective for Implementing Solutions in Education and Coaching". *Chinese American Scholars Association Journal* (<http://www.g-casa.com/conferences/berlin/papers/Galbiati.pdf>).

De Vecchi Galbiati, P. (2013). "Disruptive Innovation and Memetics: how to accelerate change and transformation in business modelling, organization and development". *Chinese American Scholar Association Journal* (<http://www.g-casa.com/conferences/slovenia/papers/Galbiati.pdf>)

De Vecchi Galbiati, P. (2014). "Augmented didactics in Kindergarten12: An Italian Case History". *Journal of Pedagogic Development, Centre for Learning Excellence, University of Bedfordshire* (<http://www.beds.ac.uk/jpd/volume-4-issue-1/augmented-didactics-in-kindergarten12-an-italian-case-history>).

De Vecchi Galbiati, P. (2014). "New frontiers for corporate education: the self-learning". *People Development Magazine* (http://peopledevelopmentmagazine.com/new-frontiers-for-corporate-education-the-self-learning/).

De Vecchi Galbiati, P. (2014). "Back to school: how to increase people engagement for company evolution". *People Development Magazine* (<http://peopledevelopmentmagazine.com/back-to-schools/>).

Folke, C., L. Pritchard, F. Berkes, J. Colding, and U. Svedin (2007). "The Problem of Fit between Ecosystems and Institutions: Ten Years Later". *Ecology and Society* 12 (1) (<http://www.ecologyandsociety.org/vol12/iss1/art30/>).

Hillel, J., H. J. Chiel, and R. D. Beer (1997). "The brain has a body: adaptive behaviour emerges from interactions of nervous system, body and environment". *View Point Review* (<http://e.guigon.free.fr/rsc/article/ChielBeer97.pdf>).

Petrovna Zamaraeva, Z. and N. Vladimirovich Pankratov (2013). "The Essence of the Category 'Adaptive Resources of Self-Sufficiency'". *Sociology, Middle-East Journal of Scientific Research* (<http://www.idosi.org/mejsr/mejsr15%2810%2913/14.pdf>).

Schönera, G., M. Doseb, and C. Engels (1999). "Dynamics of behavior: Theory and applications for autonomous robot architectures, Robotics and Autonomous Systems". *Science Direct* (<http://www.sciencedirect.com/science/article/pii/0921889095000496>).

Scott Kelso, J.A. (1997). "Dynamic patterns: The self-organization of brain and behaviour". *MIT Press* (<http://books.google.it/books/about/Dynamic_Patterns.html?id=zpjejjytkiIC&redir_esc=y>).

Smith, B. P. (2004). "The Rise of Civilization and the Evolution of Personality". *Personality Research* (<http://www.personalityresearch.org/papers/smith.html>).

Zhu, P. (2015). "The Self-Adaptive Organisms". *Digital Master Tuning XXVIIII* (<http://futureofcio.blogspot.it/2015/02/digital-master-tuning-xxviii-self.html>).

Raffaella Folgieri

Technology, Artificial Intelligence and Keynes' Utopia: A Realized Prediction?

1. Introduction

In a globalization scenario, the current economic crisis has highlighted the weaknesses of the global economic system, strongly interdependent and increasingly dependent on market fluctuations and the geopolitical balance. The difficulty of getting out of the current situation is evidenced by the numerous efforts to find systemic solutions that would be able to cope with other future manifestations of a crisis of such a magnitude. An interesting aspect is represented by the emergence of themes that form the backdrop to the economic crisis but, at the same time, they represent the most obvious manifestations. Among these economic themes, unemployment is the one that most affects the public and scholars, representing, in a system in which the consumer is the pivot of the economy and international finance, the Achilles heel of economic recovery.

But what else could be expected, given the present scenario? Are there other contributing factors that have accelerated the process of crisis? Above all, does the technological acceleration that we have been witness to in the last decades represent an opportunity or an obstacle in the search for a new system of economic balance?

Taking a cue from Keynes' predictions in his famous speech *Economic Possibilities for our Grandchildren* (Madrid, June 1930), this work analyzes the link between technological progress and employment that is the responsible for rampant unemployment, critically tackling the major themes of the economy weighed against the progress of information technology and, in particular, of Artificial Intelligence.

2. John Maynard Keynes: Economic Possibilities for Our Grandchildren

In 1930 at a conference in Madrid, John Maynard Keynes delivered a speech entitled *Economic Possibilities for our Grandchildren*, in which he addressed the issue of unemployment in a far-sighted and completely innovative way, linking it to technological advances in the present and the future, among other things. In his speech, Keynes foresaw runaway unemployment that would have occurred mainly because of technological progress, which, over time, would take over the work that used to be performed by human workers, thereby reducing the number of necessary working hours. This in turn would free up time for individuals to carry out different activities pursued not to search for remuneration but for personal growth or entertainment. Keynes' idea, which was then seen as a utopian idea, is, in fact, partially realized today. Germany, for instance, has solved the problem of unemployment in part by reducing working hours (the example of Volkswagen is significant here), realizing, in practice, the thought and the indications of Keynes. Unemployment is, today, in Italy up to 13.2%, with a peak among young people (15–24 years) of 40%. Other countries, however, are not better off.

In this speech, Keynes emphasizes that progress in general, and in technology in particular, is essential to the human aspiration to be free from wage slavery, opening a path to pursue more cultural and non-profit work. Yet Plato and Aristotle questioned the production of objects and practical activities by placing them in the background compared to the production of ideas:

> Meanwhile, they dreamed future worlds, gods and heroes using robots. *"If every instrument – fabled Aristotle – could, once ordered, work by himself, if quills could knit alone, if the bow sounded alone on the zither, entrepreneurs could do without the workers and the bosses of the slaves".* (*"The revenge part of Keynes"*, Domenico De Masi)

However, over time, a growing gap has been observed between the classes who can afford to reduce their work to a pleasant occupation and those for whom it was and is still a necessity for survival (referring to the birth of the proletariat).

Nevertheless, progress led to the surrender of the lower classes, who were able to live in degrading conditions and do exhausting work. New inventions, along with technological and industrial progress before then, gradually improved working conditions, providing a glimpse into the possibility of Keynes' envisioned utopia of total freedom from fatigue and physical work. Information technology, in recent decades, has accelerated this process, making the motto "less and less work thanks to the machines" an ideological banner that has fostered the development of technologies' and systems' optimization of time and resources. Every expert in these areas develops their own experiences, keeping in mind that information technology can help save time and labor. In contrast, however, this impulse was not met with a real reduction in the demands of work. In contrast, the less time it takes to carry out certain tasks has actually led to a greater burden for each worker. In this way, the amount of product divided by the time and manpower needed to produce it, although incremental for each product produced, generates, for a certain period of time, uncontrolled growth of the demands placed on each worker. Today this situation primarily concerns workers who have to deal, directly or indirectly, with the technology and with the use of technological means, while the other categories experience progress not as liberation from work, but as a threat to employment. Along these lines, as clearly shown by De Masi, even organizational innovation has contributed to the reduction of labor required to produce goods or services. In fact, Taylor, the father of scientific management, provides several examples of the effect of organizational innovation on employment.

All of this may seem positive at first glance, the utopic desire for liberation from the need to work, but companies have not actually redistributed the workload to reduce working hours. Instead, they tend to reduce staff, up to a collapse in the entire system, with the boomerang effect of unemployment. In fact, growing unemployment (and therefore decreasing revenue per capita) consequently decreases the consumption of goods and services, with obvious repercussions for the entire economic system, which is already precarious because of delicate geopolitical balances, as previously mentioned. Therefore, reducing working hours is a principle, not a practice. In fact, according to the believe that the recognition of merit is closely related to the extension of permanence in the work place, new employees tend to prolong their working hours.

As De Masi states: "*homo faber* prevails systematically *homo cogitans* and especially the *homo ludens*, multiply, rather than reduce, the causes of unhappiness taken as 'natural' and even as a providential opportunity atoning of living beings".

We are therefore still far from Keynes' ideas about technological unemployment and its consequences. Nevertheless, the economist had anticipated the limits of his own theory, today practiced not with the desired effects, which states that unemployment should be combated by reducing taxes and increasing investment. While the technological acceleration we are seeing today was not yet predictable in 1930, Keynes realized that technology would indeed have a key role in the economy and that it would be a significant contributory cause of unemployment.

The first words uttered by Keynes in his speech at the opening of the Madrid conference were: "Right now we are suffering from a severe attack of economic pessimism. I think this is a very wrong interpretation of what is happening". These are words that ring true with the mood of contemporary times.

Keynes begins with a historical excursus to describe all the efforts made by man to free himself from the crushing weight of hard work before proceeding to explain how technological unemployment is a necessary and transient evil to reach liberation from work.

In his speech, Keynes distinguishes between absolute needs (exhaustible) and related needs (inexhaustible), predicting that we will soon be able to meet the absolute needs to devote all our energy to non-economic purposes, thus reaching the utopian ideal that humanity has always pursued. He speculates, in fact, that there will be a first phase in which the work will decrease dramatically, which in turn will raise the need for the redistribution of employment. This phase is then followed by a second phase of a cultural nature, in which man will have the leisure time to occupy his free time with other interests. This will lead finally to the third phase in which a change in the moral code will serve as the culmination of the two transitional earlier periods.

As Keynes concludes:

> I see us free, therefore, to return to some of the most sure and certain principles of religion and traditional virtue – that avarice is a vice, that the exaction of usury is a misdemeanour, and the love of money is detestable, that those walk most truly in the paths of virtue and sane wisdom who take least thought for the morrow. We shall once more value ends above means and prefer the good to the useful. We shall honour those who can teach us how to pluck the hour and the day virtuously and well, the delightful people who are capable of taking direct enjoyment in things, the lilies of the field who toil not, neither do they spin. [...]
> of course, it will all happen gradually, not as a catastrophe. Indeed, it has already begun. The course of affairs will simply be that there will be ever larger and larger classes and groups of people from whom problems of economic necessity have been practically removed. The critical difference will be realised when this condition has become so general that the nature of one's duty to one's neighbour is changed. For it will remain reasonable to be economically purposive for others after it has ceased to be reasonable for oneself. The pace at which we can reach our destination of economic bliss will be governed by four things – our power to control population, our determination to avoid wars and civil dissensions, our willingness to entrust to science the direction of those matters which are properly the concern of science, and the rate of accumulation as fixed by the margin between our production and our consumption; of which the last will easily look after itself, given the first three. [...]

3. The Responsibilities of Artificial Intelligence

A phenomenon much discussed today is the emergence of so-called unnecessary work (Graeber 2013). Just open the pages of financial newspapers to read the growing concern surrounding this phenomenon, which is causing a deep malaise among those most affected. Technological unemployment, in fact, extends not only to industrial and production activities, but is also related to new intellectual activities, which are facilitated by easy-to-use technological tools and therefore cause the need for a smaller labor force.

Few people are willing to admit that their professional tasks could actually be carried out in significantly less time. In fact, many see the creation and maintenance of fictitious jobs at companies willing to create surplus labor market for fear of falling into a kind of social depression that could cripple the current social order, based on the combination of production-consumption.

Indeed, psychology, and especially cognitive psychology does not underestimate this issue, noting that depression is a social phenomenon linked to modern times and periods of unemployment. When there is no concept of paid activity, such as in prehistoric times, are people depressed?.

In this scenario, since progress today is primarily technological, what responsibility can be attributed to this technology and, in particular, to Artificial Intelligence (AI), that is heading increasingly in the direction of creating machines and intelligent systems that might become capable of replacing humans, even in intellectual tasks?

When Keynes delivered his vision of this utopia, it generated a heated debate within the disciplines of automation, the forerunner of today's AI. Scholars wondered whether it was ethical to study automation, and what systems and technologies could replace man in his daily activities and work. This discussion continues today, especially in light of the effects of technological progress on employment.

Taken from a historical perspective, it is obvious that technological innovation and human work evolved and continued to evolve in a symbiotic way, increasingly defining a framework in which human intervention is less and less necessary. Electronics, automation, information technology and the culmination reached with Artificial Intelligence, have meant that the pace of change taking place has become more and more accelerated, causing confusion and distress in all that part of humanity that, still immersed in the daily needs, has not embraced the utopian vision of Keynes.

ICT in general has already provided support to daily activities and work, has streamlined workloads and, in many cases, transformed or even made obsolete some professions or trades. Publishing is the sector that is suffering the most from this process (Greco et al. 2013). AI is leading to changes having an even greater impact, which often is not obvious but that now pervades every aspect of our lives. Indeed, cybernetics, bionics, robotics, neural networks (and, more generally, machine learning algorithms) have brought about significant changes in productive activities and services. The use of artificial intelligence systems has been extended from applications in scientific computing to more general domains. Algorithms and intelligent machines are used in various fields. The economic, business, medical, industrial and legal sectors are among the most obvious, but in general, any area in which predictions need to be made (i.e. by applying a sort of human intuition) using data without sufficient prior knowledge are affected by the change. One example of this is fraud detection machines and algorithms that can identify a fake check, fraud, or a scam, and lock the transaction. In the medical field, machine-learning methods are used for diagnosing and prognosing diseases, or for making efficient imaging machines that are able to detect pathologies. Automation, which is the field from which AI (cybernetics) has been playing an important role since World War II is a reality in industry today. The transport system uses AI algorithms (e.g. the state railways since the '80s) to regulate traffic. Even in the legal field, there are various applications of AI algorithms, such as the system SEL adopted by the Appeal Court of Rome, Italy, to automate the civil process, providing advice to judges to determine quickly the formal and substantial correctness of the preliminary stages of a trial. Scotland Yard uses an artificial intelligence system that can detect obvious similarities between different types of crime. Furthermore, experts in the financial markets use systems based on neural networks for making financial predictions. In the defense field, voice, writing and image recognition systems have been a reality for many years. In telecommunications, AI algorithms are used for signal cleanup (Vaseghi 2008). Weather forecasts are made using AI tools (Gardner and Dorling 1998). Search engines on the web, the systems of commerce and e-commerce are, too, based on AI algorithms (Lawrence 2000). Even in sports (Lapham and Bartlett 1995), or in the human resource selection process (Mehrabad and Brojeny 2007), these systems are used to improve process efficiency in, respectively, training and recruitment.

Today, our daily activities are based largely on Artificial Intelligence algorithms and the use of intelligent machines, though these instances may not be as obvious as the other applications mentioned previously. The wide field of AI includes, in fact, not only systems and algorithms, but also cybernetics and robotics, also more and more present in the daily life, not only in industrial automation, but also in applications such as exoskeletons (Yang et al. 2008) or robotic assistance to humans (Jacobsen et al. 2004). For instance, robot appliances, but also nurse cyborgs are already being used in some Japanese hospitals. In this socio-economic climate, Karel Čapek, who coined the term 'robot', foresaw in his play *RUR* (1920) (Čapek 2004) the possibilities inherent in the construction of artificial beings cheaper than the human being, but performing the same tasks. We have come a long way from Shakey (Nilsson 1984), the first mobile robot on wheels, which was created at the Stanford Research Institute in Menlo Park, California, during 1967–69. Think of the progress made with Asimo (Chestnutt et al. 2005).

In recent times, the excess labor force in industry was moved from manufacturing to services and, later, when it was revealed to be exuberant also in services, it was hijacked in the ICT sector (which employs 40 percent of the active population in the advanced countries). What will happen after? Will there be other emerging sectors or, rather, the utopia of Keynes? Do the times require a slowdown?

If answer to this question seems to be an obvious 'yes' because humans actually have difficulty in adapting so quickly to the new social order, then why is AI accelerating to create machines that can replace humans?

One might say that all AI experts are proponents of a Keynesian utopia, in fact, now there are neural networks that replicate the mechanisms of the human brain almost completely and very closely. We should start from the past to fully understand how strong man's urge was to create machines that can replace us in everyday life. This calls to mind Pascaline (1642) created by Blaise Pascal, a computing machine created for calculations; Babbage's wheel calculators; the creation of ENIAC[1] (1946) by Von Neumann. In contemporary AI, machines and algorithms (ANN, Artificial Neural Network, SOM, Self Organized Maps, machine learning, automatic learning) tend to replicate and replace man, imitating his cognitive processes. These efforts aimed to create thinking machines, as created by Turing in 1950. When thinking of AI, who does not think of HAL, the on-board computer in the film *2001: A Space Odyssey* by Stanley Kubrick?

The great challenge of AI is to expand the capacity of artificial systems by combining the creativity, judgment, and intuition of human intelligence, with the speed, accuracy, and attention to detail of artificial systems. Without going into the philosophical or ethical questions, it is clear that being able to create a system that brings together human and artificial abilities can result in a product usable in any context, with some guarantee of success.

But will AI actually be able to create systems and machines that could replace humans?

In the field of AI, there is an eternal debate that pits those who support the so-called strong position against supporters of the weak position. In the philosophy of AI, the strong position (strong AI, a term coined by John Searle) argues that forms of AI can be made to reason and solve problems, and demonstrate self-awareness:

> According to strong AI, the computer is not merely a tool in the study of the mind; rather, the appropriately programmed computer really is a mind. (John Searle)

On the other side of the debate, weak AI supports the use of programs to study or solve specific problems, excluding the possibility of achievement of self-awareness

---

1 The ENIAC was absorbing a large amount of electricity, so much so that its first power-up caused a blackout in the western district of Philadelphia. Like all the first samples of computers, it was very cumbersome, occupying a space of nine by thirty meters (180 square meters) and weighed about 30 tons.

for machines, which can never be defined as intelligent because they cannot actually think. About the IBM supercomputer Deep Blue[2], Drew McDermott wrote: "Saying Deep Blue doesn't really think about chess is like saying an airplane doesn't really fly because it doesn't flap its wings", arguing that Deep Blue possesses limited intelligence restricted by the size of its intellect. On the contrary, many refute this claim, arguing that Deep Blue only follows a program encoded in it.

Proponents of the weak AI argue that machines can never truly become intelligent, and therefore can never replace humans, while the supporters of strong AI believe that attaining true self-consciousness for machines will be possible in the future.

Von Herder asserts that there is no space for AI. He wonders, in fact,

> "what does it mean to think? Speak inwardly, that is expressing itself marks acquired. Speaking means think aloud, in the flow of these thoughts, much can be for us only supposed and opined; but if I think really an object, it never happens without a sign. In thinking, the soul continually creates a unit of its manifold" (Von Herder 2002).

Conversely, Dennett argues:

> Well, then how can the brain extract meaning from certain things? At what point can we talk about consciousness? These are the questions to which the cognitive sciences are trying to give an answer, trying to reduce the internal representation and those who experience the above representation of the machines. A computer can do it. The great insight of Turing was this: reduce the semantic machine to a syntactic machine. Our brains are nothing more than syntactic machines, which, however, extract meaning from the surrounding world, or work as semantic machines. We are in the presence of a paradox, but not a mystery, as many would have us believed. I do not believe in mysteries, they are only problems that we do not know how to approach. If we think we have found a mystery, we probably just misunderstood the problem. What is certain is that consciousness is less mysterious than you think: it develops from what the brain does – or how syntactic machine – and not by what it is made from. (Dennett 2006: 42).

The action of the mind unfolds, according to Searle, through intentionality, a basic property of the mind, a mental process that connects the inner world to the outside world.

> The ability of speech acts to represent objects and states of things in the world – says Searle – is an extension of the most biologically fundamental ability of the mind to relate

---

2 Deep Blue is the IBM RS/6000 computer equipped with 512 processors working in parallel and programmed to play chess. On May 11, 1997, it beat the strongest human chess player in the world, Gary Kasparov. No one argued that mankind had finally built a thinking machine, but the IBM computer had been shown to have an intelligent behavior greater than the challenger.

> the body with the world, by means of mental states such as belief and desire, and, in particular, through action and perception. (Lyons 1995)

But does the strong position of AI tend to create thinking, creative machines? There are already many attempts to do so. Among others, Lamus[3] was presented in July 2012, a super computer that composes classical music, designed and built at the University of Malaga by a research group in computational intelligence (coordinated by Eng. Francisco J. Vico), assisted by the pianist Gustavo Diaz-Jeres. The objective was to test the Turing test, and, in fact, the first concert on a computer (with the title *Can machines be creative?*) was streamed on July 2, 2012 and dedicated to Turing.

Even earlier, in 2010 in Udine, the poet and mathematician Hans Magnus Enzensberger created a machine that writes poems automatically, in complete independence[4].

There are also mini-robots developed at the University of Lausanne, which, equipped with a brain based on artificial neural networks (imagine those of an insect), use a mechanism of 'electronic reproduction', evolving according to the laws of natural selection: the robots have the best chance to reproduce, and then to combine the 'digital genome' (or the weights of the neural network of the artificial brain) with that of another sample (randomly). Within a few generations, the robots have demonstrated an increase in intelligence, the ability to find 'food' independently (locating a charger) and avoiding 'poison' (a location similar to that of the 'food' but causes deactivation of the robot). Robots have also learned how to alert the presence of food to one another using light emitters they are provided with, thus developing a kind of language.

Furthermore, in 2012 at the Massachusetts Institute of Technology laboratories, researchers developed a prototype robot that can change its shape. The project was called Milli-Motein[5] and its robots have extraordinary potential. Consider,

---

3 The first composition made by Lamus: <http://www.youtube.com/watch?v=bD7l4Kg1Rt8>.

4 The machine was manufactured by Solar SpA, Udine. Enzensberger "wanted to experience concretely the theory that has fascinated many writers, beginning with von Chamisso, poet at the turn of the eighteenth and nineteenth centuries, but especially the work of Raymond Quenau and of surrealists like Breton. The first, with its 'factory of potential literature' has shown that you can write a million of billion sonnets starting from the verses of one poem".

5 Milli-Motein is a robot similar to proteins, which naturally change their shapes. See "I robot di Losanna" (episode of SuperQuark): <http://www.rai.tv/dl/RaiTV/programmi/media/ContentItem-446dbf19-e344-401b-b7fe-67698c4daad5.html?p=0>.

for example, potential applications in domestic daily life, such as a lamp that can be transformed into a cup.

Finally, the SyNAPSE system (Systems of neuromorphic Adaptive Plastic Scalable Electronics) announced by IBM in 2009 emulates brain capacity linked to feelings, perceptions, actions, interactions and cognition. IBM's aim is to build a chip for cognitive calculating. In cooperation with a team of researchers from the Lawrence Berkeley National Lab and Stanford University, the company has built a particularly innovative simulator of calculation mechanisms, memory and communication. It is also innovative for the biological details inspired by the neuro-physiology and neuroanatomy[6].

AI efforts are, however, currently oriented towards the creation of computational models that can simulate intelligent systems (or be used by intelligent systems) to understand, but even more so to reproduce the workings of the human brain. Intelligent systems that derive from or benefit from this are the most diverse: think of agents, robots, systems in the environment, entertainment and learning applications that are already an everyday reality, tools that replace humans in their activities.

All this might be frightening, obviously, for all the reasons already discussed. But then why do humans want to create intelligent machines? Why do AI scientists pursue these goals?

It is not easy to answer these questions, and the Keynes' utopia seems the most plausible explanation: the reasons are to be found in the human aspiration to gain more free time and the brain-spiritual dimension.

Surely AI, above all other ICT disciplines, has great responsibilities in today's employment crisis and, moreover, the process has not yet been arrested: every day machines and systems are becoming not only more powerful, but also more creative. If we believe, in fact, that AI will always be in the service of intelligence and of human motivation, then it is sufficient to develop fast, flexible, powerful and mobile systems and adapt over time to the connected changes. If, on the contrary, the intention is to make the most of the opportunities to develop these machines, then artificial minds need to show human characteristics, such as creativity, mindful of the inherent risks. Therefore, technology and AI will wonder if the machines will have to feel part of a connected reality or simply be the means of connection, thus favoring, or slowing down, the process of realization of the Keynes' utopia, for which they are largely responsible.

---

6 Currently IBM has stated that the system shows the characteristic of a cerebral cortex of a cat (about 4,5% of human brain).

4. Conclusions

With regard to Keynes and his utopia, what aspects are still timely, and which have forcefully returned to the forefront? It appears that we are right in the first phase predicted by the economist. In fact, even Japanese efficiency had to surrender, due to the increase in unemployment caused by the progress of technology (think of Fujitsu which, despite the success of the company, has laid off thousands of people and announced a hiring freeze). Staff reductions, in fact, no longer concern only businesses that are in crisis. Instead, it extends to all the companies that have chosen or, rather, had to choose a high-tech model, opting for what is now called "jobless growth" or "development without employment" (Caballero and Hammour 1998). But will this model really result in greater levels of creative, richer intellectual activity, a life pervaded by aesthetics, freed from economic needs, the objective of which will be the subjective self?

Keynes' speech seems prophetic in light of what we are experiencing. It is hoped that his utopia will soon come to fruition, raising us from the overwhelming and inevitable anxieties of the middle period in which we are living. We can't say if the technology and, in particular, the AI can be of help in the solution to the current situation, or act as accelerators for the final structure advocated by Keynes. However, surely the involvement of technological disciplines makes them responsible for the pressure of events, if not for the entire process.

Obviously, AI is not yet able to totally replace human beings. Instead, systems and machines created using AI technology are becoming a valuable source of support to the human decision maker in a growing number of situations, or as a replacement for dangerous or heavy kinds of work. It goes without saying that progress in general has always led to a transformation and, given the acceleration of technological progress, even in this case technology has led to the disappearance of certain professions and trades. However, the responsibility of technological progress and AI, which is state-of-the-art technology, must be shared with social and economic practices, which thus far have not had the effects suggested by Keynes (e.g. reduction of working hours) but try to maximize profits at the expense of liberation from work as put forth in the Keynesian utopia.

Meanwhile, waiting to see if the Keynesian utopia will be realized, humans will have to live with fears about the misuse of technology, which have found expression in artistic masterpieces such as *Brave New World* (Aldous Huxley 1932), *Modern Times* (Charlie Chaplin 1936) and *1984* (George Orwell 1949), who were able to masterfully interpret fears of technological domination present in the collective imagination.

References

Caballero, R. J. and M. L. Hammour (1998). "Jobless growth: appropriability, factor substitution, and unemployment". *Carnegie-Rochester Conference Series on Public Policy*, Vol. 48: 51–94.

Čapek, K. (2004). *RUR* (Rossum's universal robots). New York: Penguin.

Chestnutt, J. M., G. Lau, J. Cheung, J. Kuffner Hodgins and T. Kanade (2005). "Footstep planning for the honda asimo humanoid". *Robotics and Automation, 2005. ICRA 2005. Proceedings of the 2005 IEEE International Conference on.* IEEE: 629–634.

Dennett, D. C. (2006). *Dove nascono le idee*. Rome: Di Renzo.

Gardner, M. W. and S. R. Dorling (1998). "Artificial neural networks (the multilayer perceptron). A review of applications in the atmospheric sciences". *Atmospheric environment* 32(14): 2627–2636.

Graeber, D. (2013). "On the phenomenon of bullshit jobs". *Strike! Magazine* 10–11. <http://strikemag.org/bullshit-jobs/>.

Greco, A. N., J. Milliot, and R. Wharton (2013). *The book publishing industry*, 3rd ed. New York: Routledge.

Jacobsen, S. C., M. Olivier, F. M. Smith, D. F. Knutti, R. T. Johnson, G. E. Colvin, and W. B. Scroggin (2004). "Research robots for applications in artificial intelligence, teleoperation and entertainment". *The International Journal of Robotics Research* 23 (4–5): 319–330.

Keynes, J. M. (1933). "Economic Possibilities for our Grandchildren". *Essays in Persuasion, Collected Writings*, Vol. IX: 358–373.

Keynes, J. M. (1997). "Economic Possibilities for our Grandchildren". *Frontier Issues in Economic Thought* 2: 343–344.

Lapham, A. C. and R. M. Bartlett (1995). "The use of artificial intelligence in the analysis of sports performance: A review of applications in human gait analysis and future directions for sports biomechanics". *Journal of Sports Sciences* 13(3): 229–237.

Lawrence, S. (2000). "Context in web search". *IEEE Data Eng. Bull* 23(3): 25–32.

Lyons, W. E. (1995). *Approaches to intentionality*. Oxford: Oxford University Press.

Mehrabad, M. S. and M. F. Brojeny (2007). "The development of an expert system for effective selection and appointment of the jobs applicants in human resource management". *Computers & Industrial Engineering* 53(2): 306–312.

Nilsson, N. J. (1984). *Shakey the robot*. Menlo Park: Sri International Artificial Intelligence Center.

Taylor, F. W. (1914). *The principles of scientific management*. New York: Harper.

Turing, A. M. (1950). "Computing machinery and intelligence". *Mind* LIX (236): 433–460.

Vaseghi, S. V. (2008). *Advanced digital signal processing and noise reduction*. Chichester: John Wiley & Sons.

Vogelheim, E. (2005). "Working Time for Managerial, Professional, and Production Jobs: The Case of Volkswagen AG Germany". In: Hegewisch, A. (ed) *Working time for Working Policies*. Washington DC: Friedrich Ebert Stiftung Occasional Papers (Friedrich Ebert Stiftung).

Von Herder, J. G. (2002). *Herder: Philosophical Writings*. Cambridge: Cambridge University Press.

Yang, C. J., J. F. Zhang, Y. Chen, Y. M. Dong, and Y. Zhang (2008). "A review of exoskeleton-type systems and their key technologies. Proceedings of the Institution of Mechanical Engineers Part C". *Journal of Mechanical Engineering Science* 222 (8): 1599–1612.

Angela Lupone

The "Green Beautiful" Option for the WTO-Multilateral Trading System: Cutting the Edge between Feasibility, Pragmatic Approach and Utopia in Governing the Multiple Aspects of Globalization in Food Markets

1. The Green Beautiful as a Possible Option for Planet Earth

Green Beautiful is the name of the marvelous utopian planet and the ideal society of the future depicted by Coline Serrau in the outstanding but unfortunately not very well-known French movie *La belle verte*, acted in, written, and directed by Serrau in 1996. On Green Beautiful, people live conscious, healthy, peaceful, simple lives in harmony with body, mind and nature, and enjoy longevity. Technology has evolved to the point that it is already overruled, even overhauled by biological knowledge and phenomenal quantal consciousness. Hence, the film is a sort of "from the future back to the origin of mankind".

Indeed, a long trip has brought humanity from the Big Bang through wars, disasters, the so-called progress of the Industrial Era and extremely stressed technological evolution to the edge of its own existence. This continues on and on in a sort of circle that has developed back (or further on) to the potentialities of human nature itself. Hence, the human mind on Green Beautiful is able to do everything we still use technology for all by itself. In fact, telepathy is a common communication tool. Through telepathy, the inhabitants of Green Beautiful can move to Earth and other planets, can 'disconnect' and drive back to awareness, presence and good will exhausted earthlings' minds. Furthermore, newborns receive and spread strength and nourishment from simply being held in one's arms.

In order to communicate through telepathy from Earth, the inhabitants of the planet use water: fountains, lakes, wash-basins, bathtubs, etc. Water is not only the source of life, but the basis of communication and relations. Thus, water becomes the common heritage of humanity that must be protected in order to ensure communication for future generations. On Green Beautiful there are 'concerts of silence' but also 'sessions of laughter'. People eat everything: fruit, vegetables, olives, kebabs of raw vegetables and grains, but no meat. Instead they eat beans. Finally, everything is agreed and settled peacefully in the yearly planet meeting held on the top of a mountain and headed by the eldest member of the community, Ozam – and when they reach the summit, people are so exhausted they don't want to quarrel anymore.

In short, *La belle verte* provides a unique and unusual look at our own world of material bondage, environmental decay, the dictatorship of money and the marketplace, inverted or diverted value perception, widespread confusion, loudness to the point that we cannot even hear our own interior voice. On the one hand, the message is encouraging: Green Beautiful is our own planet itself, just a couple thousand years in the future. On the other hand, if Green Beautiful really is a possible option, the question is how do we manage to become Green Beautiful, and how do we succeed in tackling the basic needs of planet Earth? With regard to my research interests, namely international trade law, specifically the WTO's approach to intellectual property rights, non-trade concerns and food safety issues in the view of balancing the market's rules and humanity's needs, the question becomes: have we already started out on this trajectory, and if so, do we have the necessary tools? Are we already doing something to boost individual consciousness and general environmental well-being in the globalized society without threatening the achievements reached from the multilateral trading system and through other forms of institutional market integration at the international, regional or bilateral level? In short, where are we, and where are we going?

But, let me go back to the movie and to the annual planetary meeting headed by old Ozam as mentioned previously. There is a special point on the assembly's agenda entitled "travels to other planets". Even though Green Beauty's inhabitants aim to learn from their neighbors, nobody has been to Earth for 2000 years! Surprisingly, it is the only destination, at least at the beginning of the story, without volunteers willing to travel to it. The reason for this is that Earthlings are supposedly unable to teach anything to move further forwards. From the perspective of Green Beautiful, on Earth society continued to degrade, as one tells who had previously been on Earth. This individual also believes that there is nothing that can be done for Earthlings. In this respect, it had been really difficult down there, and life on Earth is marked by the "law of the strongest, squashed women, massacres and *no distribution*". In short: "they would have needed centuries to recover". Furthermore, if one wanted something, he couldn't get it without money, even food. "But *to eat* is a *necessity*. We die if we don't eat", Mila's son says indignantly. "But they do this. If you have no money, you have nothing", is the final severe remark. In this context, it is worth noting that "necessity" is one of the main principles defining the evolution of international trade law and shapes WTO Members' commitments to health and environmental protection, as well as food safety issues[1].

The description of "Earth's Industrial Era", something Green Beautiful had passed about 3000 years before, is similarly severe. Serreau herself defines this era as "competition, literacy, mass production of useless objects, wars, nuclear technology, destruction of nature, diseases without cures, a prehistoric period". Not surprisingly for a contemporary audience, there was even a country where women had to wear veils on their faces and they did not have the right to drive a car!

Of course there is much more in the movie, but another aspect of life on Earth nowadays seen from the futuristic Green Beautiful worth mentioning is the "hierarchy" since "the men think they are superior to the women, adults are better than children, the human beings are superior to animals and plants and there are races", as an old woman explains. Viewed from Green Beautiful, Earth is a big planet, with many different continents, which results in diverse groups of humans separate from each other. When they meet, some groups think they are superior to others, which can result in mass killings. One thing that makes Green Beautiful distinct from Earth is that they have a single race, a single climate, and a single *way of development* on their little planet. Surely this makes everything easier, but it is not because Earthlings didn't have the chance: they didn't even try[2].

Here we arrived at our question of what I dare call the 'state of the art'. Does international trade and/or economic law sufficiently consider the Green Beautiful option in managing issues such as food safety and security, access to medicine, technology transfer, environmental threats, and the protection of biological diversity, animal and plant health, traditional knowledge, and human and basic rights?

Before moving on, let me stress another interesting statement in the movie. Indeed, this brings me a little further from the main topic of this article. Nonetheless, it is connected to other studies collected in this volume, specifically the issue of communication and the tools that can be used to enhance and increase it. Ozam and some other people on Green Beautiful do not want to give up the Earth. However, most of the participants at the assembly are convinced that Earthlings are not open to communication, as one young man shouts. So, the question becomes: are humans actually already like this? Are we playing with computers instead of exercising our brains? Are we really only using 10 percent of our brain's capacity? Communication plays an important role in the movie. Unfortunately, it is not

1 See Boisson De Chazournes (2010) and Mavroidis (2012).

2 It's worth noting that later in the movie it will be Ozam himself who will be defined as a racist by Mila, the protagonist.

possible to focus on this point now, but let us keep this question in mind. Certainly, there is partial overlapping between the issue of strengthening conscious societal and individual behavior with respect to safety-related matters at the international level and through international cooperation. And the issue of promoting fair and effective use of technologies and communication tools in order to avoid them becoming threats for individual rights and freedoms.

Finally it is Mila who travels to Earth. She wants to know where she comes from, as her mother was an Earthling.

2. Promoting Adequate Food Access Through Consumer Behavior and Consciousness

As already pointed out, Green Beautiful had passed through an industrial era. Perhaps surprisingly from our point of view, this period was followed by the so-called 'Era of Processes', surely an alarming scenario. Indeed, at that time all the people who produced products that damaged human health and the health of the environment had been judged as culprits of genocide and of committing crimes against the planet! Who were they? The food and chemical industries, weapon factories, tobacco and alcohol, pharmaceutical and nuclear industries, car producers, architects, and the many doctors and politicians who had become rich by allowing it to happen. This evolved into a civil war. The Time of Boycotts followed. It was the human community's ultimate weapon. The inhabitants of Green Beautiful didn't buy things anymore and threw things away that caused harm. In fact, fewer purchasers meant less power. Even the army and the police could not do anything about the boycott. It was the "chaos before renaissance". Given the historical trajectory of Green Beautiful, the question becomes: do we too, necessarily have to go through this chaos as a kind of path of purification in order to regain control of things, instead of being led by them?

Looking at the WTO-Multilateral trading system, I personally do not think that we will come to the desperate point described in the movie. In this respect, the GATT (General Agreement on Tariffs and Trade) already had tools in place at the time of its adoption. The WTO, almost through newer jurisprudence of the WTO/Dispute Settlement Body (DSB) and WTO/Appellate Body (AB), seems to be searching for more practical and sustainable solutions, since throwing out the baby with the bath water is not a viable solution[3]. In any case educating individuals on awareness and on conscious societal behavior is becoming a fundamental

3 Referring to the SPS Agreement see EC-Approval and marketing of Biotech Products, WT/DS291/R (2006); Australia-Salmon, WT/DS18/AB/R (1998); with respect to

goal of national and international governance and politics related to trade in food, access to medicines, the protection of biological diversity.

When considering food safety governance in the WTO multilateral trading system, for example, we must accept the fact that members approach the issue differently at the regulatory level. There are historical, cultural, economic, developmental, political and religious reasons for this. Nevertheless, food travels as does any other good; it passes frontiers with its qualities and its dangers. National restrictions on trade aim to combat health risks, among other goals. Nevertheless, these restrictions can easily hide or impose protectionist measures. International cooperation and cooperation among international organizations and non-governmental institutions is growing in order to avoid further food crises. On the other hand, harmonized systems of safety regulations based on general standards are being adopted or recognized by relevant international organizations, such as the WHO, Codex Alimentarius Commission, the International Office of Epizootics, various international and regional organizations operating within the framework of the International Plant Protection Convention to elaborate *ad hoc* standards or widen the acceptance by states in mutual recognition of national and private safety standards and certifications[4]. This cooperation boosts trade and international relations, building bridges and bringing states (and people) closer together.

Food quality, adequacy and safety must be protected without compromising access in less developed countries. This could mean, for instance, accepting the use of GMOs in contrast to the level defined by high national standards of protection. In this context, it is up to the international community to set, if necessary, new *fora*, rules, or procedures for the protection of general values and needs at stake, including quality, traditional farming and combating situations that threaten agriculture such as land grabbing, which threatens biodiversity, traditional knowledge and agricultural methods, and the rights of local populations.

An important question is whether this kind of balancing among different interests and non-economic concerns connected to trade should be the principal task of the existing WTO-Multilateral trading system in a world moving quickly towards emerging forms of plurilateralism through Regional and/or Preferential Trade Agreements (RTAs, PTAs) aimed at enforcing, at different levels, WTO-plus or WTO-extra regulations[5]. It is a challenging goal I would personally welcome.

---

the TBT Agreement see US-Tuna II, WT/DS/381/R (2012); US-Clove cigarettes, WT/DS406/AB/R (2012).

4 Even though there is already concern about the growing power of certification agencies. On recognition of technical and safety standards see Howse (2011).

5 Hufbauer and Cimino Isaacs (2015).

Furthermore, it should include related aspects of the action against international corruption and international organized crime connected to trade in food and medicine.

But to return to awareness as a tool to boost sustainable and science-friendly development that could help avoid something like the so-called Boycott Era in Green Beautiful, I cannot resist repeating that educating about conscious societal and individual behaviors must remain, (or become) the fundamental goal of national and international governance and politics, especially if related to safety, trade in food, access to medicine and the protection of biological diversity. Incidentally, this also means promoting a commitment to protect, occupy, or even conquer the right to exercise one's freedom of choice in light of adverse or "bad" market forces. Simply *banning* access to unsafe food or GMOs from a foreign country, for instance, is no longer enough. Instead, there is a need for civil society to demand overall protection of biodiversity and traditional farming practices at the international and transnational level. With regard to these points, consumer behavior is already positively evolving: production processes that endanger human, animal and plant life are often prohibited by laws or at least condemned and boycotted by consumers. In many states and in the European Union, production and distribution processes are almost (or supposed to be) traced and products' features clearly marked. It goes without saying that governments, companies, and groups are influenced by civil society's pressure and mood, which is becoming increasingly relevant and effective, but also dangerous for power players, given the easy and open access to social networks.

From another perspective, poorer countries have also gained stronger consciousness with respect to their own needs and opportunities. All of this is influencing economic relations and the future role of the multilateral trading system. Even though the steps forward do not yet appear to be remarkable, they are proving that we are moving in the right direction. Nevertheless, we still have a long way to go, and civil society could play a stronger role in confirming that unsafe, spoiled, counterfeit, insufficient, undiversified, and inadequate food and foodstuffs are real threats to trade. Subsequently, the challenges the international community and the transnational society will first have to face in order to neutralize the potential danger of all this are not bans or restrictions to trade, but continuing to develop and spread methods and tools for positive integration for safety issues among members. If we look at the WTO Agreement on Sanitary and Phytosanitary Measures (SPS) and its latest interpretation by the Dispute Settlement Body (DSB) and the Appellate Body (AB), there are opportunities for a better future[6].

---

6 For relevant jurisprudence see *supra*, footnote 3.

3. Trying to Avoid the "Boycott Era": WTO Tools for Sustainable Balance in the Food Trade and Other Safety Issues

Starting from the multilateral trading system first set by the GATT in 1947 and developed further during the Uruguay round (1986–1994), let us take a brief look at the main threats we are being confronted with, along with some legal tools already available for enforcing food safety in international trade law.

First of all, are food safety requirements an obstacle to free trade in the multilateral and market-oriented trading system? Indeed, the core of the current debate on food safety and international trade might be summarized with the following questions:

- How do different national food safety regulations interact in our interdependent world?
- Should the WTO consider these food safety regulations as simply barriers to trade? If so, should the aim be to reduce or abolish them in one way or another? Or should the "national level of protection" be considered a *unicum* that needs or even deserves to be preserved and protected[7]?
- How does one get around this dilemma?

With regard to the specific significance that the issue assumes in international trade in general, is it currently possible to assert that national food safety requirements which somehow refer to internationally accepted parameters (such as the widely-held acceptance of a general right to 'adequate' food) prevail – limiting them – over those of international trade in the free trade context? Considering some of the latest developments in WTO practices and jurisprudence, I tend to (carefully) say yes.

In particular, measures *necessary* to protect human, animal or plant life on the basis of *national level of protection* adopted by a given member, are presumed to be compatible with GATT. However, "if not applied in a manner which would constitute a means of arbitrary or unjustifiable discrimination between countries where the same conditions prevail or a disguised restriction on international trade" (Art. XX Chapeau GATT).

As a matter of fact, and exactly through Art. XX, Chapeau trade requirements have for the most part prevailed in the comparative evaluation of interests at stake,

---

7 Along these lines, see Echols (2008), Raustiala (2008) and Scott (2008), with particular reference to the role of the Committee on Sanitary and Phytosanitary Measures and the WTO Dispute Settlement Body.

as confirmed by the practice of the WTO/Appellate Body. Even though correct from the strict regulatory perspective, unfortunately this approach does not take into sufficient consideration the point of view set forth in *Green Beautiful.* In other words, a food safety issue is relevant insofar as it is trade related. Consequently, an issue is relevant insofar as it is trade related. This approach first requires checking if, and within which limits, food safety today is considered trade-related by the WTO system itself, and if so, under which conditions can an internal restrictive measure affecting international trade[8] be justified when it comes to food safety matters?

Indeed, considering the harsh economic repercussions recorded in cross-border trade relations and the consequences on the international and internal judicial level linked to the latest scandals and food disasters, the link between food safety and well-balanced international trade cannot be denied. Though this relationship between cause and effect is generally acknowledged, when it is a matter of specifically evaluating the compatibility of a given restrictive national measure on international trade in WTO law, the members' standpoints prove to be very distant from each other. This also applies for cases in which safety standards agree upon common means of operation to be established. And this both from the North-South perspective (developed countries' positions in respect to the ones of developing countries with reference mainly to the sustainability of measures asserted by developing countries), as well in the North-North relationship among peers/developed countries. One simply needs to go back to the hormone-treated meat case[9] or to point to the divergent positions of the United States and the European Union on GMOs and recall the ongoing discussion in the EU around the supposed negative effects of the potential Transatlantic Trade and Investments Partnership (TTIP) with the USA, particularly in the food sector.

Given the fact that the multilateral trading system is aimed above all at hindering protectionist practices by members, there are fortunately several rules within the GATT (General Agreement on Tariffs and Trade), GAS (General Agreement on Services) and TRIPS (Agreement on Trade Related Intellectual Property Rights), i.e. limitations on trade in goods, import restrictions or other prohibitions linked to food safety that may justify an exception to general principles.

---

8 Reference made to decisions listed in footnote 3.

9 WT/DS26/R/USA *EC Measures Concerning Meat and Meat Products (Hormones).* Complaint by the United States. Report of the Panel, August 18, 1997; WT/DS26/AB/R – WT/DS48/AB/R *EC Measures Concerning Meat and Meat Products (Hormones). Report of the Appellate Body,* January 16, 1998: Adinolfi, G. 2004, La soluzione delle controversie in Venturini, G. (ed) L'Organizzazione mondiale del commercio, seconda edizione: Milano, Giuffré Editore, 231–241.

For example, this is true whenever there is a danger to health or the need to intervene in combat practices that may mislead the consumer (Art. XX lit. b) and lit. d), General Exceptions); whenever there is a risk of food scarcity within a member state or quality control must be ensured (Art. XI.2 lit. a), lit. b), lit. c), General elimination of quantitative restrictions); and whenever the packaging of goods may be misleading or fraudulent (Art. IX.2, Marks of origin). In general, restrictive measures are temporary (Art. XI.2b) and must be carried out in compliance using the principle of non-discrimination (Art. XIII).

In the adoption and implementation of national measures, transparency and publicity are to be guaranteed (Art. X. 2), and the principle of necessity and coherence of the measure must be complied with. Finally, scientific evidence attesting to the appropriateness of the trade-restrictive measure must be provided. With regard to quality, reference should be made above all to the TRIP's Agreement in addition to Article IX of the GATT. In any case, this all falls within a broader context aimed at understanding the role of the WTO in relation to issues which, though not always covered by the WTO multilateral treaties, are already debated in the WTO Committees as they have an impact on the international balance of economic interests and basic human needs within the WTO. Other measures are set out by the SPS Agreement (Agreement on Sanitary and Phytosanitary Measures), in the Agreement on Agriculture, as well as in the Agreement on Technical Barriers to Trade (TBT) and in the Agreement on Import Licensing Procedures. The discipline established under the SPS Agreement is particularly relevant for the purposes of reducing trade constraints by way of approximating national legislations. Here it is worth mentioning that the harmonization mechanisms are established under Art. 3 and the provisions relevant to equivalence and mutual recognition appear in Art. 4. As already noted above, international standard-setting bodies, whose role within the multilateral trading system has been redesigned by the principle of prevalence of international law, acquire special relevance when it comes to harmonizing objectives. The Agreement grants these bodies a sort of indirect power of attorney, which has raised some criticism in connection with the asserted legitimacy deficit of such institutions.

With respect to intellectual property and the aim of balancing economic interests and fundamental needs, reference should be made to the international regulation of certain intellectual property rights, such as geographical indications, patents, plant species and brands, as well as to the impact of domestic and international trademark exhaustion regimes and to the growing phenomenon of trade in counterfeit food. Indeed, the rules governing geographical indications (intended as indicators of food quality) have implications for food safety (Arts

22–24 TRIPs). Negotiations within the Doha round and so-called TRIPs-plus and TRIPs and regional or bilateral agreements substantially strengthened and extended international protections, even though the primary purpose of this was not safety[10]. Another new aspect relevant to food safety is represented by the relationship between the TRIPs Agreement and the international regime of the previously mentioned and so-called triplet: biotechnology, biodiversity and traditional knowledge[11]. When dealing with international trade law on food safety, we cannot avoid taking into consideration newer trends that go beyond the WTO, such as ongoing TTIP-negotiations or other already concluded RTAs and PTAs[12]. Apart from the harsh question of the WTO/DSB's jurisdiction with respect to RTAs/PTAs, which is very uncertain and controversial, most of the regulations at stake could well be subject to future evolutionary interpretations and applications by the WTO bodies, potentially contributing to the harmonization of standards and to the acknowledgement of policies (international, regional or even national ones) connected with food safety on an international level. Symbolic of this is the action taken by the SPS Committee with reference to the Codex Alimentarius[13].

4. Facing Food Safety Complexity at The Doha Round

The Doha Ministerial Declaration dated November 14, 2001 already reaffirmed the joint commitment of all WTO members to emphasize some collective interests that go beyond the pure aim of trade liberalization, including the promotion of sustainable development, environmental conservation and healthcare, specifically food safety in future multilateral negotiations. WTO members thus

> recognize that under WTO rules no country should be prevented from taking measures for the protection of human, animal or plant life or health, or of the environment at the levels it considers appropriate, subject to the requirement that they are not applied in a manner which would constitute a means of arbitrary or unjustifiable discrimination between countries where the same conditions prevail, or a disguised restriction on international trade, and are otherwise in accordance with the provisions of the WTO Agreements. (para. 6)

---

10 See e.g. Free Trade Agreement between the European Union and its Member States, of the one part, and the Republic of Korea, of the other part (signed on October 6, 2010, provisionally applied since July 1, 2011). For more information on TRIPs-plus agreements see: Grosse Ruse-Khan (2011), Seuba (2013) and Roffe / Escudero / Seuba (2015).

11 See Taubman / Wager / Watal (2012).

12 See literature cited in *supra* at footnote 8.

13 On the subject, see Scott (2007).

At the same time, members reaffirmed the obligation, within negotiations relating to the WTO Agreement on Agriculture, to ensure that the special, differentiated treatment for developing countries allows these countries to cope with their needs, including food safety (para. 13), as well as the importance of implementing and interpreting the TRIPs Agreement in a manner supporting public health (para. 17).

In any event, the perspective of the Doha Declaration with regard to future negotiations strengthens what has already been implied in some of the covered agreements. Reference may be made, although to a limited extent, in the very Preamble of the WTO Agreement, which defines the general objectives of the Organization (in particular, the sustainability of development and environmental conservation), to the general exceptions of the GATT Agreement 1994 (Art. XX) and of the GATS Agreement (Art. XIV) that makes it possible to waive the duties of trade liberalization for the purposes of protecting the basic needs of society. Furthermore, the Preamble to the Agreement on Sanitary and Phytosanitary Measures (SPS Agreement) affirms members' desire to "improve the human health, animal health and phytosanitary situation in all Members". Moreover, as far as the basic principles of the TRIPs Agreement are concerned, members may adopt measures that are necessary for protecting the food supply and public health (Art. 8).

Unfortunately, several obstacles to the development and implementation of this perspective have appeared[14]. Furthermore, the Doha round started more than fifteen years ago and is currently in stalemate. Since a conclusion of the Round is not likely in the short term, the means of protecting food safety are promoted mainly in harmonization efforts driven by relevant international organizations dealing with standardization processes, bilateral governmental cooperation and voluntary recognition of the equivalence of certifications.

14 Indeed, the WTO Dispute Settlement Body, though proving to be attentive to issues related to the protection of health, operates within the narrow limits of interpretation of what is set out in the agreements. See for example, in the WTO Appellate Body's decision *EC-Hormones* WT/DS26/AB/R, *European Communities – Measures Concerning Meat and Meat Products (Hormones)*, Report of the Appellate Body, January 16, 1998, where it was stated, in reference to the precautionary approach established by Article 5(7) of the SPS Agreement, that such a rule "has not been written into the SPS Agreement as a ground for justifying SPS measures that are otherwise inconsistent with the obligations of Members set out in particular provisions of that Agreement" (para. 124).

5. The Issue of Qualification, Its Implications for International Food Safety Governance and Final Remarks

There is one issue that remains at the forefront due to its far-reaching implications: food safety qualifications, which includes product quality. Thus, international trade law tools with reference to safe food mean not only identifying multilevel protection measures in forms of international, regional or bilateral co-operations that safeguard against unsafe, spoiled or counterfeit food, but also mechanisms based on common criteria that facilitate the international monitoring of the use of harmful substances contained in food. It also means checking the authenticity and conformity of the quality and origin indications. Furthermore it implies promoting the mutual recognition of standards and national certifications in the context of limiting their restrictive effects on international trade. The more the observer probes into these mechanisms, the more he or she becomes aware of the many issues at stake, particularly the economic and political interests involved. It becomes evident that health dangers often stem from political and economic choices; the mandates of several international organizations overlap and the variety of ethical and cultural values held by different regions and populations play a fundamental role. But this is not all. Issues like the relationship between food safety and international organized crime recently caused headlines in Italy, in other European countries and in the US, which is particularly alarming[15]. Last but not least, one should not overlook the debate regarding the use of biotechnology in food production.

In order to prevent misunderstandings, we must always bear in mind that at the international level, the concept of food safety refers to two different situations: the right to (access) food, on the one hand, and the right to safe food and food quality on the other[16]. There is no doubt that there are points in common, with intersections and some overlap between the two areas, so that for the most part it is not possible to deal with the former without also taking the latter into consideration[17]. And there is no doubt that the latest contemporary practice in international cooperation and relations takes the issue of food safety and quality as one of the cornerstones of international governance strategies, also relating to

15 See for example Saviano (2012).
16 The English language uses the name of food security in the first case and food safety in the second one. There is extensive doctrine on this point. See Moyo (2007); Hospes / van Dijk / van der Meulen (2010); Breman and Termeer (2010); Szajkowska (2010).
17 Hospes / van Dijk / van der Meulen (2010).

the right to food and the right to health in general[18]. On the subject of healthcare and food quality in particular, several other important matters converge, especially if food security is also intended as the 'adequacy' of food 'in relation' to the basic needs and requirements of a given human, societal or local group.

In addition to the aforementioned issues that are often regulated or otherwise addressed by international law and intergovernmental cooperation, there are other, more general or specific points of view. Consequently, this implies (the need for) coordination of actions and cooperation among international organizations in many fields to promote development, environmental conservation, and to prevent and deal with man-made disasters including wars and humanitarian crises, safeguard traditional knowledge and biodiversity, as well as ensuring access to technology and, finally, respecting cultural, traditional and ethical or religious values linked to food.

In fact, food safety is a political, economic and/or cultural matter. Yet it is a 'global' one. As such, it needs to be 'governed' and not necessarily 'regulated' by international cooperation. This can no longer be accomplished simply by using traditional tools of such cooperation. Given the particular nature of food safety, it is increasingly (even within and by international organizations themselves) being addressed using other informal instruments of a more political nature, through mediation, and with soft law. These tools have been shown to be more capable than traditional ones involving and making private parties interact with institutional ones on a local level and in the international arena. And so, on the one hand we find private parties engaged in production and distribution processes, and on the other hand, institutional actors, lobbies and NGOs aim to protect individual interests of a more private nature at the global level, as for example consumer protection associations[19].

Using the example of food safety, I highlighted how the numerous issues at stake intersect, which has led some international organizations to extend *de facto* their sphere of activity well beyond the original boundaries, especially in well-known food crises, such as mad cow disease, bird flu and the poultry dioxin scandal[20]. The repercussions from these emergencies crossed national borders, with serious consequences both for private operators and global economic growth. Consequently, the direct link between healthcare, food quality, and international

18 Johnson (2007); Dunoff (2011).

19 See Raustiala (2008).

20 In this sense, the international economic institutions themselves e.g. the International Monetary Fund and the World Bank, have become increasingly involved in healthcare, food safety and food quality.

trade has come to light[21]. This helps explain why we cannot avoid drawing attention to the relationship between food safety and the (evolving) WTO trading system[22].

Moreover, it is still a largely unexplored area that reveals interesting potential[23]. Significant is the role of the WTO's Committee on Sanitary and Phytosanitary Measures (SPS Committee) in relation to the international harmonization of safety standards by way of cooperation with (and incorporation of) Codex Alimentarius and NGOs to fill what was defined as the "global gap between private standards and international cooperation"[24]. Similarly, I think this can also be said for the TRIPs Agreement and its implications for food health issues. Other intersections may also come to light with time in relation to the actions promoted in various 'sensitive' sectors by the World Intellectual Property Organization (WIPO).

Particularly in respect to TRIPs, while the matter of labeling origin is more clear, at least concerning the existing link between the name of origin and a specific quality of the product, an issue which goes far beyond the economic right of exclusive use of the designation of origin itself, other potential TRIPs issues and/or situations are arising that need to be 'governed'. These adhere, as already mentioned above, to the protection of biodiversity, traditional knowledge and to the 'exclusion' of these from being patented. The debate on the limits of patent rights gives rise to further arguments that are variously supported by international practices. It outlines in a new and heterogeneous way the confrontation among basic human needs between local governments and corporations on the one hand, and the 'glocal' human community in a wide sense on the other. The fact is that the discussion is set far away from the traditional North-South approach. In this context, the issues connected with technology transfer entailed by the production of safe and adequate food, as well as with checking safety levels are important to remember. Technical assistance in the checking procedures by richer countries on behalf of developing ones plays an important role too. Finally, there is a need for straight coordination with WIPO's activity, particularly regarding specific objectives in developing countries.

Lastly, I would like to draw attention to a further aspect of food which has not really been taken into consideration so far, but which, in my opinion, is becoming

---

21 It also touches on trade in services in a limited sense. For more on the WTO's GATS Agreement and with reference to the migration of doctors and nurses from the African continent, see Gathi (2010); spec. Aginam, (2010).

22 See Hufbauer and Cimino-Isaacs (2015).

23 In this sense, with reference to the WTO's SPS Committee, see Scott (2008).

24 See Scott (2008: 263).

particularly worrisome with respect to food market governance in the broadest sense. I am referring to the presence of large corporations in the international market, which dominate food production as a whole[25]. Indeed, just a few multinational companies are the main operators in the agricultural and food industries in the international arena. The maximization of profit, which is any company's priority, tends to clash with the societal necessity to safeguard the general public's most fundamental needs. Biological and food diversity do not necessarily coincide with profit, since the pursuit of profit might entail risks for food safety. We should therefore ask ourselves what role (if any) the WTO or other institutions should play when it comes to safeguarding food safety and ultimately human and environmental health. In fact, notwithstanding the investments and fair competition regimes, as of today, they are not an integral part of WTO regulations.

Indeed, the WTO seems to have assumed a role that goes beyond its primary goal of hindering protectionism and liberalizing international trade. The WTO has reverted to debating non-trade matters, such as food safety. As it currently stands, do necessary food safety requirements relating to internationally accepted parameters take precedence over trade requirements, even within the context of the WTO? There are encouraging signs that we are moving in this direction, where one will be able to say that the obstacle to the balanced performance of trade within the WTO is the 'absence' of food safety, rather than national standards of protection, particularly if mutually agreed upon or internationally recognized.

At the beginning of this study, I posed the question of whether international trade law is evolving towards the "Green Beautiful Option", or in the opposite direction, towards the tremendous "Era of Boycotts and Massacres" described in the movie. Taking an optimistic view, the answer to the question of whether we are headed towards the first option is "sort of, but not yet sufficiently", and "for sure not" with respect to the other scenario.

References

Aginam, O. (2010). "'Predatory Globalization': The WTO Agreement on Trade in Services, Migration, and Public Health in Africa". *ASIL Proceedings*: 139–146.

Boisson De Chazournes, K. (2010). "Standards et norms techniques dans l'ordre juridique contemporain : quelques reflexions". In: Boisson De Chazournes, K. (ed) *International Law and the Quest for its Implementation*. Leiden: Liber Amicorum Vera Gowlland-Debbas.

25 See Raj (2007); Magrini, (2012).

Breman, G. and C. Termeer (2010). "From Food Security to Food Quality: spreading standards, eroding trust?" In: Hospes, O. and I. Hadiprayitno (eds) *Governing Food Security. Law, Politics and the Right to Food*. The Netherlands: Wageningen Academic Publishers: 185–227.

Dunoff, J. L. (2011). "New Voices I: Global Health, Trade and Common Resource Regimes. Introductory Remarks". *Harmony and Dissonance in International Law. ASIL Proceedings*: 95–96.

Echols, M. A. (2008). "Remarks". *ASIL Proceedings*: 259–268.

Gathi, J. T. (2010). "Hot Topics in GATS and Human Rights. Introduction". *ASIL Proceedings*: 127–128.

Grosse Ruse-Khan, H. (2011). "Protecting Intellectual BITs, FTAs and TRIPs: Conflicting regimes or mutual coherence?" In: Miles, K. and C. Brown (eds) *Evolution in Investment Treaty Law and Arbitration*. Cambridge: Cambridge University Press: 485–515.

Hospes, O., H. van Dijk, and B. van der Meulen (2010). "Introduction". In: Hospes, O. and I. Hadiprayitno (eds) *Governing Food Security. Law, Politics and the Right to Food*. The Netherlands: Wageningen Academic Publishers: 19–38.

Howse, R. (2011). "A new device for creating International Legal Normativity: The WTO Technical Barriers to Trade Agreement and 'International standards'". In: Joerges, C. and E.-U. Petersmann (eds) *Constitutionalism, Multilevel Governance and International Economic Law*. Oxford: Hart Publishing: 383–395.

Hufbauer, G. C. and C. Cimino-Isaacs (2015). "How will TPP and TTIP Change the WTO System?" In: *Journal of Intellectual Economic Law*: 1–18.

Johnson, C. D. (2007). "What Future for the Doha Development Agenda and the Multilateral Regime?" *ASIL Proceedings*: 243–244.

Magrini, M. (19 February 2012). "Le misteriose regine del trading". *Il Sole 24 Ore*.

Mavroidis, P. C. (2012). *Trade in Goods: The GATT and the other WTO Agreements regulating trade in Goods*. Oxford: Oxford University Press.

Moyo, D. (2007). "The Future of Food: Elements of Integrated Food Security Strategy for South Africa and Food Security Status in Africa". *ASIL Proceedings*: 103–108.

Raj, P. (2007). *Stuffed and Starved*. London: Melville House.

Raustiala, K. (2008). "The New Politics of Regulatory Cooperation: The case of Food Safety. Introductory Remarks". *ASIL Proceedings*: 259–268.

Roffe, P., S. Escudero, and X. Seuba (2015). "From TRIPs to Preferential Trade Agreements, Including the Trans-Pacific Partnership Agreement and related Trends in the European Union: Challenges for Emerging Countries".

International Institute for Sustainable Development, <http://www.ipekpp.com/admin/upload_files/Report_3_54_From_2237283020.pdf>.

Scott, J. (2007). *The WTO Agreement on Sanitary and Phytosanitary Measures. A Commentary*. Oxford: Oxford University Press.

Scott, J. (2008). "Remarks". *ASIL Proceedings*: 259–268.

Saviano, R. (23 July 2012). "Dalla carne alla mozzarella. Camorra Food Spa serve a tavola". *La Repubblica*.

Seuba, X. (2013). "Intellectual Property in Preferential Trade Agreements: What Treaties, what content". *The Journal of World Intellectual Property Law*: 240–261.

Szajkowska, A. (2010). "Food safety governance from a European perspective: risk assessment and non-scientific factors in EU multi-level regulation". In: Hospes, O. and Hadiprayitno, I. (eds) *Governing Food Security. Law, Politics and the Right to Food*. The Netherlands: Wageningen Academic Publishers: 201–227.

Taubman, A., H. Wager, and J. Watal (eds) (2012). *A Handbook on the WTO TRIPs Agreement*. Cambridge: Cambridge University Press.

Gloria Regonini

Administrative Simplification Between Utopia And Nightmare

1. A Common Nightmare: Bureaucracy

Almost a century ago, Max Weber described the competitive advantage of bureaucracy over previous forms of organization of public agencies with the following words: "Precision, speed, unambiguity, knowledge of the files, continuity, discretion, unity, strict subordination, reduction of friction and of material and personal costs – These are raised to the optimum point in the strictly bureaucratic administration" (Weber 1925d, 1978: 973). On the grounds of such vision, Weber believed that the efficiency and self-referential nature of the bureaucratic machine foreclosed a source of risk for representative institutions, incapable of exerting effective control over its activities.

Today, delays and inefficiency of public agencies feed into the feelings of dissatisfaction common to many citizens in a wide range of countries differing in terms of their institutional settings, organizational cultures and political balance. Bureaucracy has been blamed for many of the failures of public intervention, and those who complain about its inadequacy range from Nobel Prize[1] winning scientists to D-Day[2] veterans, as well as entrepreneurs and families, taxpayers and recipients of welfare subsidies.

In Italy, discontent stemming from lengthy and complex administrative procedures is widespread. From international adoptions to certificates attesting major illnesses, from scholarships to tax returns, from stadium building projects to fishing, there is no area of human activity that is not perceived as burdened by the unbearable weight of red tape.

International comparisons broadly confirm one fact: Italian opinion of public administration is not the result of a collective prejudice, but is based upon solid objective evidence of serious inefficiencies. According to the Worldwide Governance Indicators (WGI) published by the World Bank between 1996 and 2014, Italy

1 "Nobel winners say scientific discovery 'virtually impossible' due to funding bureaucracy," Rebecca Smith, *The Telegraph*, June 2, 2014.

2 "Veteran's anger as French bureaucracy threatens to derail 70th anniversary of D-Day", *Sunday Express*, April 21, 2014.

performed consistently below average among high-income OECD countries with respect to all governance dimensions. Particularly significant was the deviation in the values of those indicators, which are most directly related to the functioning of public administration: government effectiveness, regulatory quality, and control of corruption (World Bank 2015).

2. A Common Dream: Simplification

If bureaucracy can be regarded as a nightmare, simplification is the dream that politicians, regardless of their political orientation, constantly promote as the ideal model of good governance for a healthy relationship between citizens and public administrations.

The goal of 'zero bureaucracy' holds considerable charm, as politicians, journalists and bloggers know very well. The battle between the evil of bureaucracy and the sake of simplification is often described with military or religious metaphors (e.g. an anti-bureaucracy 'task force', 'war', 'ban', 'vision', 'mission', 'crusade'). The promise is that simplification can be similarly achieved by simple means: the deed can be done by just cutting, deleting, or getting rid of regulations, offices, or procedures.

The Italian case clearly indicates that the relationship between means and ends is not so linear. Twenty years of simplification initiatives have revealed how the dream can transform itself into a new nightmare, producing the reverse of what was expected. In fact, hundreds of measures taken by governments on opposite ends of the political spectrum – involving different governmental levels, appointing simplification ministers and town councilors and depleting considerable human and financial resources – have produced a layering of rules that have generated enormous costs for policymakers, without conferring any substantial benefit.

3. The Usefulness of Extreme Cases

In many ways, Italy's relationship with paperwork has atypical features, making it an extreme case: "The Italian legislative corpus has long represented a labyrinth for even the shrewdest legal practitioner because of its complexity and sheer volume" (Borghetto and Visco 2015: 106). Even among Southern European bureaucracies, Italy is notable for the persistence of its legalistic and formalistic administrative cultures (Galanti 2011).

Yet the unexpected and unintended effects in this field are not exclusively an Italian anomaly. That the utopia of simplification can be transformed into its opposite is not a remote hypothesis. Indeed, in radical libertarian theory, this is an inevitable outcome, as stated by David Graeber with his 'iron law of liberalism': "Any market force, any government initiative intended to reduce red tape and promote market forces will have the ultimate effect, increasing the total number of regulations, the total amount of paperwork, and the total number of bureaucrats the government employs" (Graeber 2015: 7).

The empirical evidence reported in international comparisons of the bureaucratic burden leads to more cautious evaluations. If it is true that no nation is completely immune to this problem, the opinions of citizens and entrepreneurs in terms of their relationships with their nation's administrations vary greatly from country to country. Italy consistently places below the average in measures of regulatory quality, a finding that can be used as a magnifying glass to better examine the dynamics that lead to worse outcomes in order to tackle them more effectively: "The 'generalizability' of case studies can be increased by strategic selection of critical cases [.] Atypical or extreme cases often reveal more information because they activate more actors and more basic mechanisms in the situation studied" (Flyvbjerg 2001: 77–78).

The case study examined in the following pages relates to all the measures implemented in Italy to reduce the administrative burden on businesses.

4. Simplification for Businesses

Since the 1990s, and even more so after the economic crisis of 2008, Italian entrepreneurs have been complaining about the unbearable burden of bureaucratic procedures that affect each phase of their activities, from starting a business, to its possible developments or its termination

In surveys conducted in recent years to gauge the opinions of business leaders within the Global Competitiveness Reports of the World Economic Forum, the 'Inefficient government bureaucracy' item almost always ranks first among 'the most problematic factors in doing business' in Italy.

Complaints of entrepreneurs often carry a tone of exasperation due to the sheer amount, complexity and slowness of required administrative formalities. In 2014, for example, Antonello Montante, legality delegate of Confindustria, declared: "Bureaucracy causes more damage than the Mafia"[3].

3 La Repubblica, Palermo edition, February 25, 2014. Antonello Montante, who is also president of Confindustria Sicilia, is under investigation for accusations of Mafia related activities as of January 2016.

These criticisms have had a strong influence on the decisions made by political institutions. Moreover, in the last decade, the simplification of administrative procedures for businesses, especially for small and medium enterprises (SME), has become an explicit, continuous and non-controversial objective, even for the European Union. With the Small Business Act (SBA) of 2008, the Commission set the following objectives for itself: "Design rules according to the 'Think Small First' principle" and "Make public administrations responsive to SMEs' need[s]"[4]. These principles, together with the finding that the little progress made by some countries, including Italy, are regularly reaffirmed in the annual European Competitiveness Reports.

On a national level, Italian governments, regardless of political orientation, have adopted numerous measures since 2008, such as draft-laws approved by Parliament, executive orders, rules of procedures, codes, and three-year plans, all of which were aimed at easing the administrative burden placed on Italian companies. Political leaders have learned to give attractive 'nicknames' to these laws: "Masters in our own home" (2001), "Cutting administrative burdens" (2008), "Development Decree" (2011), "Simplify Italy" (2012), and the "Decree of doing" (2013).

During that same period, all twenty Italian regions have implemented a number of agreements signed with the central government and have produced laws and simplification decrees on important matters within their jurisdictions relating to trade, agriculture, forestry, fisheries, the environment, health and social services, urban planning, construction, health, and civil protection.

If the outcome of the government's actions were measured by the number of words approved by means of formal deliberations, the simplification for SMEs 'Italian Style' would deserve an extremely positive assessment. But it is not so. The main objective of simplification is to make it easier to open and run businesses in Italy. With respect to this aim, the most recent international comparisons confirm the persistence of a strong delay. The Doing Business 2016 report (World Bank 2016) places Italy among the lowest-ranking European countries. The evaluation is particularly low for operations that directly involve the public administration and the judiciary power, such as paying taxes and enforcing contracts. The Global Opportunity Index 2015 by the Milken Institute (Wickramarachi and Savar 2015) places Italy in the same negative position with respect to the ability to attract investments, with a continuous drop in performance since 2009. Even according

4 Commission of the European Communities 2008. Communication from the Commission to the Council, the European Parliament, the European Economic and Social Committee and the Committee of the Regions – "Think Small First" – A "Small Business Act" for Europe, <http://eur-lex.europa.eu/legal-content/EN/TXT/PDF/?uri=CELEX:52008DC0394&from=EN>.

to the Global Competitiveness Report 2015–2016 by the World Economic Forum, while Italy has recovered some positions compared with previous measures, productivity remains low, "as a result of long-standing constraints such as burdensome red tape (139th) and labor market inefficiency (126th)" (World Economic Forum 2015: 30). Nevertheless, although government bureaucracy is confirmed as the most problematic factor for doing business in all the advanced economies, Italy's position is still clearly far behind the average.

After more than a decade of simplifications, this paradoxical result has been recognized in documents published by the Italian government itself: "The long and uncertain time, the excessive costs and the large number of obligations continue to represent a major obstacle for those who want to 'do business'. Over the years, regulatory announcements and measures have followed one another, yet they have failed to effectively address this crucial issue for the country's growth and development"[5].

In fact, the never-ending process of making laws and subsidiary legislation has generated perverse effects. The continuous layering of interventions concerning planning consents, environmental safety, and company registrations for tax and social security purposes, causes standards to overlap and intersect with each other, increasing uncertainty among the very public officials who should enforce them, and among the citizens who have to comply with them.

5. A Perverse Problem

In some respects, the fact that the regulation of SMEs is a never-ending process can be regarded as normal, as technologies change, the economy changes, and their impact on the environment as well as on working conditions and products themselves also changes.

But the Italian case pushes itself far beyond these progressive adjustments. A cross-section of the initiatives stratification can be found under "company simplification" on the Ministry for Simplification and Public Administration's website[6]. The 229 web pages listed represent a catalog of broken dreams of various successive governments, with their 'cutting-laws', 'new agendas', public consultations, and memos 'containing explanatory guidelines'. Pages and pages are repeated with the same objectives, without a logical connection, no milestones (and often no dates), were it

5 Governo Italiano 2015. *Agenda per la semplificazione 2015–2017, Rapporto di monitoraggio agosto 2015*, <http://www.italiasemplice.gov.it/versione-testuale-news/agenda-impresa/>.

6 <http://www.funzionepubblica.gov.it/home.aspx (accessed on January 28, 2016)>.

not for some reference to the decrees, reported in the form of unreadable PDF files[7], as if no one will ever bother to check up on their promises to see if they were kept.

In other words, SMEs comprise one of the sectors currently facing greater complications in legislation.

This significant growth in complexity produces the very opposite effects of those desired. First of all, public officials find it very complicated to keep abreast of the constant changes, to review the forms and use databases built under now-obsolete laws.

For the recipients of these interventions, compliance is difficult, even for those with the of best intentions. As anyone can see in following specialized blogs, engineers spend more time studying decrees than new building materials.

For the past several years at the end of each December, the sitting government sends the Parliament a bill called the *Milleproroghe* [A Thousand Deferrals] also in official documents, and which delays the enforcement of new and old rules for six or twelve months. This established practice is clear proof of the 'Italian Style' regulation difficulties.

The final result is the implosion of the very meaning of the word law, as recognized in the ruling of an administrative court: "The sequence of cascading referrals [...] makes the *voluntas legis*[8] almost unfathomable".

This contributes to undermining trust in institutions and the civic spirit of loyal cooperation in a country that consistently ranks very low in international surveys with respect to these aspects of political culture.

6. Two Changes in Perspective

When a problem is so extensive and persistent, it is often technically labeled as 'wicked' (Conklin 2005; Grint 2005)[9], despite attempts made to resolve it. Wicked problems have vague definitions, many interrelated causes, and different features

7 For instance, an essential document like the General Directive for administrative action and the management of the Office for the 2013 government program. <http://www.governo.it/AmministrazioneTrasparente/Performance/direttive/2013/direttiva_programma_governo_2013.pdf> and <http://www.funzionepubblica.gov.it/media/1118913/decreto_unita_%2029_aprile_2013.pdf>.

8 Ruling N. 01946/2014 REG.PROV.COLL. N. 00336/2014 REG.RIC. <https://www.giustizia-amministrativa.it/cdsintra/wcm/idc/groups/public/documents/document/mdax/otc3/~edisp/yplijz2rj6nc6vckwdp2soltre.html>.

9 Australian Government – Australian Public Service Commission 2007. *Tackling Wicked Problems A Public Policy Perspective*, <http://www.apsc.gov.au/publications-and-media/archive/publications-archive/tackling-wicked-problems>.

in different contexts. Solution attempts require the integration of many skills and often produce patchy, nonlinear effects with a mix of positive consequences and negative externalities.

In any case, when facing a wicked problem one should try to change the perspective. In 2013, a mouse conquered the web with an enviable number of views thanks to a video taken as he was trying to steal a cracker[10]. To do that, he had to climb onto a step. The video records him while tries in vain for dozens of times to push the cracker onto the next step, in reality just managing to fall back on himself. At one point, the little mouse interrupts his vain efforts and jumps on the step without the cracker.

This move allows him to observe the matter from above, from a new perspective. Once back down, the little mouse builds on what he has seen and, with some minor adjustments, he finds the right position to push the crackers on the step, and then he flees with his prize.

In the following pages, we will try to apply the example of the mouse to administrative simplification, proposing two dramatic changes in perspective. The first involves a shift from a legal approach to a policy one, analyzing the problem using the knowledge provided by social and behavioral sciences. The second change of perspective implies the reversal of the opinion that simplification may be the product of mere reduction or elimination of procedures. To really simplify, one must have an extraordinary ability to deal with the unavoidable complexity of adjustment in our open societies.

7. From Laws to Policies

The Italian fiasco in the field of simplification underscores one fact: the reduction of laws through new laws has the same credibility of Baron Munchausen's pulling himself out of a mire by his own hair. A real improvement of the relationship between citizens and administrations will not result merely from a legal logic, albeit graciously concerned with its own self-containment. Official documents have actually come to this conclusion: "As it is acknowledged, in the past, simplification measures were entrusted mainly to laws. Poor attention, if none at all, was paid to their practical application. The result is that, as citizens and enterprises know too well, many of the announced simplifications remained 'on paper'"[11]. When the entire horizon of regulation remains within the boundaries of the legal perspective, and the production of formal laws is considered the only form of

10 <https://www.youtube.com/watch?v=Uc1PbkdqRno>

11 Governo Italiano, note 6.

intervention, the very meaning of the law vanishes, becomes undetermined and loses the precise connotation that could come from the comparison with other epistemological perspectives, which provide a different interpretation of what institutions do.

In the following pages, we will try to place the simplification in another semantic space, in another 'finite province of shared meaning' (Schutz 1962): one that revolves around the concept of public policy. First of all, this entails linking the choices, whether formal or informal, explicit or implicit, of different actors in and outside institutions, using a collective problem as a common thread. According to the definition provided by James Anderson, public policy is "a purposive course of action followed by an actor or set of actors in dealing with a problem or matter of concern" (1984: 3). Secondly, this perspective requires the collection and use of the knowledge produced by social and behavioral sciences through the observation of real actors: politicians, but also street-level bureaucrats (Lipsky 1980), specialists, and the recipients of these interventions. This breakthrough, which began in the United States over 50 years ago, has an important consequence for this present study: without close observation of the actual behavior of the regulators and those they regulate, it is virtually impossible to develop any simplification proposals that will get through 'the road test'.

But applying this analytical frame to the Italian case and analyzing the war on bureaucracy as a case of (re)regulatory policy means adopting a totally different perspective from the categories in which the rulers and the ruled in this country interpret their objectives and their choices. Italian, like all the Romance languages, has a single term *politica* to define the two spheres of action that the English language calls 'politics' and 'policy'. In the practical use, the first meaning absorbs the second. When the term *politica* is accompanied by the indication of a specific public sector intervention (e.g. education, pensions), it refers to all of the laws on a particular topic. The official website of the Italian government for monitoring the implementation of the executive program only mentions "law proposals and delegated legislation" approved by the Cabinet[12]. One would search in vain through the website for documents that describe the internal logic of the various measures, their expected results, the actual stage of implementation, and any impact assessment or evaluation of the results obtained.

12 Governo Italiano, *Monitoraggio dell'attuazione*, <http://www.programmagoverno.gov.it/monitoraggio-dellattuazione-programma/monitoraggio-su-attuazione/> (accessed 27 January 2016).

In other words, the policy analytical capacity (Howlett 2009) of Italian government institutions is not only very low (Bertelsmann Stiftung 2015), but is also a skill often unknown to decision-makers themselves.

Switching from a legal perspective to a policy one means drawing upon a much wider range of governmental tools and broadening the range of factors that may influence their effectiveness: "Regulatory policy is the framework within which rules and alternatives to rules are considered, evaluated and implemented by government and public authorities seeking to influence the behavior of private actors in the economy" (Lunn 2014: 21).

If we begin by considering the possible tools of government, even in this case the family of words used in the Italian language creates important mismatches of meaning when compared with English. For instance, the Italian term *regolazione* is not the simple equivalent of 'regulation', because the former is absorbed as a synonym by the term *regolamentazione*, the formal definition of the tiny details of laws, which unlike 'regulation', leaves no room for the use of soft instruments, such as interpretation, position statements, and guidelines. 'Regulation' can be discretionary, participatory, or negotiated. *Regolamentazione* can only be top-down, command and control.

Secondly, adopting a policy perspective means broadening the range of factors that should be considered to explain the poor performance and to improve the effectiveness of interventions. The first striking fact that emerges from the reconstruction of the decision-making processes of the past decade is that the simplification arena is crowded with many actors, from the European level to the national, to regional and local ones. The venues that gather actors (boards, technical committees, task forces, etc.) keep changing their names, composition, and responsibilities with a form of institutional Keynesianism: instead of digging holes and then filling them up, organisms are first created and then eliminated, but only to be recreated under a different name. Established in 2005, the Parliamentary Commission for simplification itself also changed its name and responsibilities in 2009.

8. The Network

If we look at the most influential actors, we will note that a large part of the relevant decisions are made within "iron quadrangles"[13] who have at their corners politicians with a strong presence of SMEs in their councils, organizations representing the interests of various categories, administrative executives called

13 With reference to Griffith's (1939) 'iron triangles' among the congressional committees, the bureaucracy, and interest groups.

to enforce the laws, and professionals, who, for profit, act as middlemen between those who are regulated and the public administration and politicians.

In the analysis of the relationship between politicians and public administration, it is necessary to distinguish between the dynamics that emerge in the public communication arena and those that develop in institutional decision-making bodies. The ambiguity of the term 'bureaucracy', which can be seen both as a group of people that make up an untouchable privileged class or a set of complex and onerous procedures, represents the link between these two arenas. When politicians are (or try to be) in the spotlight of old and new media, bureaucracy is singled out as a caste that hinders the implementation of innovations decided by those who govern[14]. This game of scapegoating and blame avoidance (Weaver 1986) is vital in a country whose levels of public trust in politicians are the lowest in Europe, with a continuously negative trend[15]. When they instead hold legislative or executive roles in parliament or in the ministerial offices, politicians hardly pay attention to the linearity and formal clarity of the decisions they make, as evidenced by the bleak opinions of the Legislation Committee of the Chamber of Deputies.

Moreover, this ambiguity also characterizes the questions that organizations promoting the interests of SMEs address to the politicians with whom they are closest. On the one hand, the representatives of these groups complain about the unbearable weight of red tape. On the other hand, the regulation process is invoked, even in its most baroque forms when it comes to defending its market position threatened by foreign investments, especially in the field of trade, when trying to safeguard the products protected by quality marks, or when deregulation threatens to disrupt established business routines. The old Italian saying 'the law is interpreted for friends, and applies to enemies' has become 'simplification is for friends, red tape for competitors'.

9. Administrative Logic and Risk Aversion

The analysis of the role of administrative officers in charge of applying detailed rules and standards highlights the gap between the legal approach and the policy approach, which follows a social and behavioral perspective. The first approach is only measured

14 The Prime Minister Matteo Renzi declared before an audience of entrepreneurs: "We need to take up a violent war against bureaucracy. I'm using the word violent because we have no alternatives" (11th April 2014).

15 CESifo-Database for Institutional Comparisons in Europe, *World Economic Forum, The Global Competitiveness Index Historical Dataset 2005–2014*, <https://www.cesifo-group.de/ifoHome/facts/DICE/Values/Political-Values/Politics-and-Policy/public-trust-politicians/fileBinary/public-trust-politicians.xls>.

in one way: the degree of compliance of acts with the prescriptions set forth in long and complex legal provisions. The second approach has allowed us to understand the multiplicity of the tensions and contradictions that officials at different levels have to manage or at least absorb for many decades, namely since Herbert Simon's fundamental work *Administrative Behavior: A Study of Decision Making Processes in Administrative Organizations (1947)*. Considering public officials not as followers of a procedure for the enforcement of laws, but as actors in the implementation process, has important consequences both for the descriptive prescriptive levels. Regulatory policies for SMEs are aimed at influencing the behavior of individuals and businesses to protect public goods such as competition, the quality of the environment, and workplace safety. Their implementation is effective as long as administrative burdens are proportionate to the gravity of the feared harm, and applied appropriately and selectively, taking into account the past behavior of those who are regulated, and aim to shift from detailed preventive authorizations to a prompt identification of violations.

This ability to articulate distinct modes of regulatory implementation is quite utopian in the context of Italian public administration. First, the regulatory inputs that spill over public offices have formal features, as we have already pointed out, which are totally incompatible with their selective application and, in this case, smart. The number and frequency of laws, cross references, and the way in which the texts are written, are so jumbled that it is impossible to understand which are the purposes and which the means, which are the primary objectives and which the secondary ones. The typical language of a simplification law contains formulations like: "subsections 3 and 4 of Article 8 shall be repealed…paragraph 1a of Article 31 is repealed". Often, the changes relate to laws passed just a few months previously. In this situation, the only possible strategy for survival is the *ipse dixit*, "he himself said it", and that is the servile dependence on the interpretation memos written by superiors.

In Italy, in fact, there is a lack of consideration of how bureaucratic burdens take a toll on public administrations, which are called upon to implement laws, including simplification ones. The false assumption is the total flexibility and absorption capacity on the part of public services. However, in other countries the bureaucratic costs for public servants themselves are monitored and reported within tolerable limits. An interesting example is the 2007 British initiative 'Cutting Bureaucracy for Our Public Services', which sought "to reduce the amount of unnecessary bureaucracy faced by frontline public sector workers"[16] through the bottom-up identification of the 'key irritants' to which they are exposed.

---

16 National Audit Office (NAO) 2012, *Reducing bureaucracy for public sector frontline staff: Briefing for the House of Commons Regulatory Reform Committee*, <http://www.

The perverse complexity of regulatory inputs affecting the activity of civil servants has two serious consequences. The first is the absence of effective transparency in procedures, which creates a corruption-prone environment[17]. As is well known, Italy consistently holds a very low position in international rankings in terms of its ability to fight corruption. In a country with a strong presence of organized crime, this is not a minor problem. The second consequence is less obvious and is related to the fact that the same extreme complexity of laws ordinary citizens are confronted with also concerns the legal framework of the civil service. For an official-type, the principle "a form (or a stamp or a signature) more is better than one less" is a self-defense strategy, because it means less risk of being caught or blamed by superiors and control authorities in the case of checks. In fact, in the case of an investigation of a public servant's responsibilities, the proportionality and appropriateness of criteria would be of little or no importance. Neither are there incentives or protections for those who venture down this road.

The usual reference to the principal-agent theory as an explanation of the bureaucrats' behavior focuses on moral hazard as a source of inefficiency and lack of responsiveness. In the Italian case, this model succumbs to an alternative explanation in many situations. In 1964, Victor Thompson convincingly linked bureaupathology to the sense of personal insecurity inducing officials to give an abnormal priority to self-defense over the objectives of the organization which they work for, displaying behaviors such as "close supervision; failure to delegate; emphasis on regulations, quantitative norms, precedents, and the accumulation of paper to evidence compliance; cold aloofness; insistence on office protocol; fear of innovation; or restriction of communication" (Thompson 1964: 100). Although convictions for breaches of public laws affect a small percentage of employees, nevertheless, the stress, length and economic costs of these proceedings are so high that extreme risk aversion becomes the rational choice.

10. The Hidden Requirement of Simplification

The fourth important group of actors in regulatory policies for SMEs are intermediaries, i.e. professionals who get paid to replace the entrepreneur in interactions with administrative offices and deal with required formalities on their behalf in

---

nao.org.uk/wp-content/uploads/2012/10/0910_reducing_bureaucracy.pdf>. The NAO acts as a watchdog for the British parliament.

17 In 2015, Italy ranked second-to-last among other EU members. See Transparency International 2016, *Corruption Perceptions Index 2015*, <https://www.transparency.org/cpi2015>.

the fields of taxation, social security, urban planning, public health, etc. In Italy, these intermediaries are widespread figures. A behavioral approach allows us to understand the ambiguities and contradictions implicit in this position, even linking it to other similar professions. In the area of justice, for example, Italy, together with Luxembourg and Greece, is the country with the highest number of lawyers per capita and the length of legal proceedings is consistently higher than in other countries (CEPEJ 2014).

Even in the case of intermediaries acting in the name of the SMEs, objectively speaking, this thriving business has opposing interests to those pursued by the simplification policies, which aim to make it easy for the user to access the administration directly and autonomously. An important confirmation of this conflict came from the president of the National Institute of Social Security (INPS), which handles almost all the national pension and welfare services. In a letter to a major national newspaper, the president Tito Boeri wrote: "[...] not a day passes now without my receiving a letter from some representatives of labor consultants that threaten retaliation against INPS. My fault? Having declared at a public meeting that [...] Companies must no longer necessarily resort to intermediaries, career counselors and tax advisors and should instead interact directly with us, reducing the costs for businesses"[18].

This action has uncovered an underdeveloped aspect of our analysis of the four main actors in the regulatory policies for SMEs. Despite the frictions and mutual accusations used by the media, the elements of this network ultimately support and legitimize each other, because a change in efficiency would create problems for everyone, including the users themselves. Plans for simplification tend to systematically ignore one fact: post-bureaucratic administration does not demand fewer skills from all employees and policy makers, but rather different and more specific skills, such as being able to access channels of digital interaction; understanding English terms that increasingly often refer to the conceptual horizon of the Common Law and not to the more familiar continental Civil Law; and taking responsibility for choices that can easily get through ex ante checks, but are then subjected to more onerous ex post verifications. These are all innovations that require considerable knowledge and skills. If we compare these skills with those used by small business owners, who are loyal to the old routine based on political patronage, face-to-face relationships in administrative offices, or carte blanche delegation to intermediaries, simplification appears to be more demanding than the traditional red tape.

---

18 Tito Boeri, *Non opporsi alle innovazioni*, Il Sole 24 Ore, 17 October 2015.

This finding is of particular importance in Italy because our country has a very significant delay in the sector of adult skills. According to the 2014 OECD Adult Skills Survey (PIAAC):

In Italy, the mean proficiency scores of 16–65-year-olds in literacy and numeracy are significantly below the average of the countries participating in the Survey of Adult Skills (PIAAC). In literacy and numeracy, the younger adult population (16–24-year-olds) scores significantly below the average of the OECD countries participating in the Survey [...]. In Italy, 26.9 percent of the adult population (16–65 year olds) reports no prior experience with computers or lacks very basic computer skills[19]. To fully reap the benefits of simplification, users need significant expertise, which is quite scarce in Italy, and the situation is not any better if we consider the skills administrations must possess in order to create physical and digital communication websites that ensure both a pleasant user experience and provide effective solutions for those who access them. Unfortunately, the necessary skills and cultural understanding to build channels of this type are almost completely absent in the Italian public administration because of the generally older age of its employees, which is caused by the contraction of turnover due to the high public debt. This situation perpetuates the predominance of legal training, which until a few years ago was a necessary and sufficient condition for recruitment. This indirect adverse selection isolates the Italian administration from the technologies and communication styles used by younger generations and cross-border users.

An example of this deficiency is the troubled implementation of the Points of Single Contact (PSCs), a European Commission project launched together with the 2006 Services Directive: "They should become a single contact point where SMEs can easily obtain information, submit applications and collect decisions or other replies without having to deal with a multitude of authorities at different administrative levels, as is the case today" ("Points of Single Contact: Doing business made easier"[20]).

In June 2015, the European Commission commissioned a research study to assess "the performance of the PSCs in the 28 EU member states and three EEA member states against the PSC Charter criteria"[21]. The evaluation was not

---

19 OECD 2014. *Italy. Adult skills (Survey of Adult Skills, PIAAC)*, <http://gpseducation.oecd.org/CountryProfile?primaryCountry=ITA&treshold=10&topic=AS>.

20 <http://ec.europa.eu/internal_market/eu-go/docs/psc_en.pdf>

21 European Commission 2015. The Performance of the Points of Single Contact. An Assessment against the PSC Charter. Luxembourg: Publications Office of the European Union.

independent, having been entrusted to Capgemini Consulting and Eurochambres: Chambers of Commerce are found in many countries, including Italy, and are the main actors in the design and operation of the PSC[22]. Although the assessment of implementers-evaluators has on the whole placed the Italian PSC 'business in a day'[23] above the European average in terms of formal delegation of functions, the usability of the website is still considered disappointing when verified bottom-up: "Mystery shoppers indicated that the PSC is non-intuitive and difficult to navigate. The ease of use of the portal currently does not match customers' expectations. This refers, among others, to the extent that activities were integrated (necessity to use many different websites), the experience of technical difficulties and the extent to which users were confident that they were doing the right thing"[24].

11. Front Office, Back Office and Political Dividend

> One of the most important contributions to the breadth of literature on digital government is the identification of the fundamental difference between the so-called back and front offices: Back Office: This is the overall term for all processes and areas in a business enterprise or public authority which are carried out in the background for the citizen or customer and are not directly visible. It includes the internal processing of applications and queries received by the administration. [...] Front Office: By contrast with the back office, the processes, which take place in the front office, are processes in public administrative transactions that are visible to the customer. The further steps of processing the customer's request then take place in the back office of the administration (Anttiroiko and Mälkiä 2007: 1553)

This distinction is crucial in an era in which websites represent the most visible face of institutions. Today, the first judgment of a public policy is made on the basis of its digital identity. Emblematic from this point of view is the story of the Obama administration's website for the Affordable Care Act, HealthCare.gov, launched on October 1, 2013. When the website crashed in the first days, many commentators defined the problem not as a technical one, but as the failure of the reform itself. On October 20, 2013, President Barack Obama addressed the matter, publicly admitting that there was "no excuse". Once the technical problem had been solved, in April 2014, the Secretary of the Health and Human Services announced her resignation.

---

22 Chambers of Commerce often rely on Capgemini consultants.

23 <https://www.impresainungiorno.gov.it/>

24 *Italy. Performance of the Point of Single Contact*, (29/06/2015), <http://ec.europa.eu/DocsRoom/documents/11138/attachments/1/translations/en/renditions/nativ>

Even the policy goal to reduce red tape for businesses must first be translated into "a transformation in the business experience of frontline regulation" as mentioned in one presentation by the Better Regulation Delivery Office UK[25].

But the simplicity of the front office implies a large and complicated number of tasks for the back office. The greater the difference between what end users see and what is processed in the back office, the greater the chances of a satisfying experience for the user. Unfortunately in Italy, public administrations continue to bend the needs of the front office against the back office ones. To escape this logic, a great effort in observing actual user behavior must be made to empirically verify which cognitive maps they use and which shortcuts they take when accessing a website to interact with a public office. User strategies and expectations are a stronger objective constraint than laws and codes in determining the success or failure of the regulation, because these factors are not influenced by government decrees, but by the user experience that citizens consolidate surfing on global websites to shop online or to download films. Ignoring this fact is like ignoring the force of gravity in the engineering design of a public building.

Unfortunately, the persistent work of the back office does not pay out political dividends in terms of visibility, and it potentially eliminates one of the central actors in the policies for SMEs: the intermediaries, who in turn have strong links with interest organizations and politicians themselves. In a country where 'making a law' is considered the natural solution to every collective problem, and a large part of the political show is occupied by the tensions that accompany the law-making process, while very scarce resources are dedicated to the evaluation of its real effect, this is not a cost-effective strategy.

12. Conclusions

If used as a magnifying glass to detect problems and tensions in other nations, the reconstruction of the Italian regulatory policies for SMEs points to two important conclusions. First of all, the objective of simplifying administrative procedures is of paramount importance in influencing the degree of public confidence in institutions and the competitiveness of the national economic system. Secondly, the contribution of social and behavioral sciences brings to light problems and solutions that neither jurisprudence nor the classical economic theories of regulation can provide.

---

25 Russell, Graham 2012. *Enforcement and Inspections. A Journey Towards Change. BRDO and the UK context*, <http://www.oecd.org/gov/regulatory-policy/S1_UK_A%20journey%20towards%20change.pdf>

Since the 2008 publication of *Nudge: Improving decisions about health, wealth, and happiness* by Richard Thaler and Cass Sunstein, the interplay between the objective of simplification and the institutional use of behavioral sciences has taken on a well-known name: nudge. While this volume has been able to highlight the contribution that behavioral sciences can provide to regulation in both descriptive and prescriptive terms, however, this potential has been compressed to some extent and under-used because of an almost exclusively psychological approach to the dynamics that affect the individual choices of those who are regulated.

In this essay, I have instead tried to show that there are other analytical perspectives that can use the knowledge provided from observations of the actual behavior of the regulated and the regulators. Studying governments and the governed not from a juridical logic, but from a policy one, entails recognizing the strong influence authors such as Charles Merriam and Harold Lasswell have impressed on this discipline. The difference here, compared with fields such as behavioral economics and psychology, is that the actors are considered in their social dimension, since they are immersed in a web of relationships without which the criteria to evaluate the profitability of the alternatives they face would not emerge (Wildavsky 1987; Dunn and Kelly 1992).

The contribution that comes from a policy perspective on the issue of regulatory simplification may seem paradoxical: the first step in simplifying is to recognize the tremendous complexity involved in identifying and managing an acceptable balance between the needs of the individuals, their rights, their preferences, and the needs of the society that these same individuals belong to.

Attempts to bypass this uncomfortable but realistic conclusion have obvious limitations. In Italy, the path of simplification as a strategy of subtraction and getting rid of rules has been pursued with determination. On March 23, 2010, a real stake sealed this strategy: "Over 375,000 laws and repealed regulations were burnt at the stake. Literally. Armed with ax, pick and blowtorch the Minister for Simplification Roberto Calderoli set fire to a huge wall of boxes made up of all the provisions repealed through the work of his ministry"[26]. After six years, however, no one has ever noticed a reduction in the bureaucratic burden that can be traced to this gesture.

A second shortcut consists in understanding simplification as deregulation by other means. This strategy has often resulted in stopping the search for the difficult balance between public assets and private interests, giving more importance to the latter. In many cases, this results in shifting complications to other arenas,

---

26 Corriere della sera, *Calderoli brucia le leggi abrogate*, March 24, 2010.

such as by increasingly appealing to courts to settle disputes between rights that nevertheless are entitled to formal protection: for example, the freedom of trade, but also the right to health.

A third way is the one explored by simplification through the use of 'nudges' (Sunstein 2013). As previously pointed out, the libertarian paternalism that inspires it can easily lead recipients to make choices without a clear understanding of the overall costs and benefits implicit in the proposals to which they subscribe (Bubb and Pildes 2014).

All of these three paths are likely to lead only to a superficial simplification, moving away from the indirect effects of these strategies on the difficult political, social and economic balances upon which our civil coexistence lies.

In short, trying to keep it simple is not only complicated[27], it is also complex (Boisot 2006; Geyer and Cairney 2015).

References

Anderson, J. E. (1984). *Public Policymaking: An Introduction*. 2nd ed. Boston: Houghton Mifflin.

Anttiroiko, A.-V. and M. Mälkiä (eds.) (2007). *Encyclopedia of Digital Government*. Hershey, PA: Idea Group Reference.

Bertelsmann Stiftung (2015). *Policy Performance and Governance Capacities in the OECD and EU. Sustainable Governance Indicators 2015*. <http://www.sgi-network.org/docs/2015/basics/SGI2015_Overview.pdf>.

Boisot, M. (2006). "Moving to the edge of chaos: bureaucracy, IT and the challenge of complexity". *Journal of Information Technology* 21(4): 239–248.

Borghetto, E. and F. Visco (2015). "Governing by revising: a study on post-enactment policy change in Italy". In: Conti, N. and F. Marangoni (eds.) *The Challenge of Coalition Government: The Italian Case*. Abingdon: Routledge: 106–127.

Bubb, R. and R. H. Pildes (2014). "How Behavioural Economics Trims Its Sails and Why". *Harvard Law Review* 127: 1593–1678.

CEPEJ 2014. Report on "European judicial systems – Edition 2014 (2012 data): efficiency and quality of justice". <http://www.coe.int/t/dghl/cooperation/cepej/evaluation/2014/Rapport_2014_en.pdf>

Conklin, J. (2005). *Dialog Mapping: Building Shared Understanding of Wicked Problems*. New York: John Wiley.

27 Roz Strachan, *Trying to keep it simple? It's complicated*, May 28, 2014, Digital Marketplace Blog, Cabinet Office, Government Digital Service, <https://gds.blog.gov.uk/2014/05/28/trying-to-keep-it-simple-its-complicated/>.

Dunn, W. N. and R. Mae Kelly (eds) (1992). *Advances* in *Policy Studies Since 1950. Policy Studies Review Annual Volume 10.* New Brunswick and London: Transaction Publishers.

Flyvbjerg, B. (2001). *Making Social Science Matter: why social enquiry fails and how it can succeed again.* Cambridge: Cambridge University Press.

Galanti, M. T. (2011). "Is Italian Bureaucracy Exceptional? Comparing the Quality of Southern European Public Administrations". *Bulletin of Italian Politics* 3 (1): 5–33.

Geyer, R. and P. Cairney (2015). *Handbook on Complexity and Public Policy.* Cheltenham: Edward Elgar.

Graeber, D. (2015). *The Utopia of Rules: On Technology, Stupidity, and the Secret Joys of Bureaucracy.* New York and London: Melville House.

Griffith, E. F. (1939). *The Impasse of Democracy.* New York: Harrison-Hilton.

Grint, K. (2005). "Problems, Problems, Problems: The Social Construction of Leadership". *Human Relations* 58 (11): 1467–1494.

Howlett, M. (2009). "Policy Analytical Capacity and Evidence-Based Policy-Making: Lessons from Canada". *Canadian Public Administration* 52(2): 153–175.

Lipsky, M. (1980). *Street-Level Bureaucracy: Dilemmas of the Individual in Public Services.* New York: Russell Sage Foundation.

Lunn, P. (2014). *Regulatory Policy and Behavioural Economics.* Paris: OECD Publishing.

Schutz, A. (1962). *Collected Papers.* The Hague: Martinus Nyhoff Publisher.

Simon, H. A. (1947). *Administrative Behavior: A Study of Decision Making Processes in Administrative Organizations.* New York: Macmillan.

Sunstein, C. R. (2013). *Simpler: The Future of Government.* New York: Simon & Schuster.

Thaler, R. H. and R. Cass Sunstein (2008). *Nudge: Improving decisions about health, wealth, and happiness.* New Haven: Yale University Press.

Thompson, V. A. (1964). "Administrative Objectives for Development Administration". *Administrative Science Quarterly* 9(1): 91–108.

Weaver, K. R. (1986). "The Politics of Blame Avoidance". *Journal of Public Policy* 6(4): 371–398.

Weber, M. (1978 [1925]). *Economy and Society: An Outline of Interpretive Sociology.* Berkeley: University of California Press.

Wickramarachi, H. and S. Keith (2015). *Global Opportunity Index 2015.* Santa Monica: Milken Institute.

Wildavsky, A. (1987). “Choosing Preferences by Constructing Institutions: A Cultural Theory of Preference Formation”. *American Political Science Review* 81(1): 3–22.

World Bank (2015). *Worldwide Governance Indicators. Country Data Report for Italy*. Washington, DC: World Bank Group.

World Bank (2016). *Doing Business 2016: Measuring Regulatory Quality and Efficiency*. Washington, DC: World Bank Group.

World Economic Forum (2015). *The Global Competitiveness Report 2015–2016: Full Data Edition*. Geneva: World Economic Forum.

3. Education and Communication

Marina Brambilla–Valentina Crestani

Online University Presentations in German: Virtual, Utopian and Green. A Multimodal Analysis[1]

1. Introduction

In our paper, we propose a multimodal analysis of texts that belong to the genre of online presentations. We focus our attention specifically on university presentations as forms of institutional communication, which follows the model 'one-to-many'. The institution presents itself to students and their families with the main purpose of informing them about available educational offerings. On the one hand, these presentations have an informative function; on the other hand, they also have a persuasive function in which utopian ideals are used as strategic elements to communicate effectively and to reach readers emotionally. By underlining values like open-mindedness, along with attention to young talent and merit, the communication conveys positive feelings like happiness, interest and hope to its readers. We refer to the utopian side of the university communication as *green utopia*, a concept related to the ecology-based idea of sustainability.

In the first part of this essay, we explain what we mean by using the concept *green utopia* in the academic world. The second chapter is devoted to the linguistic terminology adopted in our paper, referring in particular to the text as an *inter*-text and multimedia text. In the third chapter, we present the main results of our analysis, which was conducted on the online presentation of the University of Duisburg-Essen in Germany – one of the youngest institutions of higher learning in Germany. In its verbal and graphic components, this presentation underlines ideals like supporting young talent and equal treatment, with concrete actions undertaken. As will be explained, this website is an interesting object of analysis to demonstrate how institutional communication can apparently create an ideal world based on principles of meritocracy. In order to support our thesis about

---

1 This paper has been written according to common ideas of the two authors. Paragraph 1 has been written in cooperation. Paragraphs 2.2 and 3.1 were written by Marina Brambilla. Paragraphs 2, 2.1, 2.3, 3.2, 3.3, 4, 4.1, 4.1.1, 4.1.2, 4.1.3, 4.2, 4.3 and 5 were written by Valentina Crestani.

utopian communication, we use the method of the multimodal analysis, which ensures consideration of:

- Many modes of communication (e.g. speech, gesture, and symbols).
- Communication as an interrelated object of analysis, which is realized in the form of texts with specific communicative functions.

The last chapter contains our final considerations.

2. *Green Utopia* in the academic world

The concept of 'green' and the concept of 'utopia' may assume indefinite contours according to different contexts in which they occur. We use this expression with a specific meaning related to the university as an institution of research, teaching and growth. We will begin the analysis with a consideration of the meaning of the color adjective *green*, after which we propose our definition of the noun *utopia*, and finally, of the idea of *green utopia* in the educational world.

2.1 Green

The word *green* includes various denotative and connotative meanings and it encodes many perceptual and cultural aspects[2]. The semantic content of *green* depends on:

A. The context of occurrence;
B. The producer of the text;
C. The recipient of the text.

In our paper, we explain these three factors as following:

A. Context of Occurrence: *green* refers to the specialized fields of education offered at German Universities, of the architecture used in university-buildings and of nutrition related to meals sold in the student cafeterias. In the first usage, it relates to the widely used concept of sustainability, as will be shown in the next paragraphs. In the second usage, *green* is connected with the new architectural principles and rules governing the construction of new buildings and the renovation of old ones in recent years. These innovative criteria were established at the European level with the introduction of a series of indicators to build new houses or renovate old constructions (see the 2010 *Directive on Energy Performance of Buildings*[3]). With reference to the third usage in the field of nutrition,

---

2 See Agnello (2013: 29–31).

3 <http://eur-lex.europa.eu/legal-content/EN/TXT/?uri=CELEX:32010L0031>

great attention is given these days to sustainable food; *green* is to be understood as food that is both good for human health and friendly to the environment, from production to the final consumption. Green criteria are important at a political-social level. European citizens are increasingly interested in sustainability, including topics like bioarchitecture, sustainable food and sustainable agriculture. Citizens are also paying more attention to the adoption of green criteria by institutions, industries, hotels, supermarkets, gyms, schools and in general all organizations that provide services to the community or sell products (see *Nielsen Global Survey of corporate social responsibility and sustainability*[4]). Educational institutions that make use of different types of buildings (e.g. universities, libraries, canteens) are an interesting field of analysis because:

1. Such institutions are of relevance to a large group of citizens. Their main target consists of teenagers, young adults and their families, who have to be made explicitly aware of the importance of sustainability.

2. As educational institutions for the citizens, they have to communicate with them in order to present their educational offerings, their philosophy, the buildings where students attend lessons, eat at lunchtime, live during the semesters, etc. Many institutes present themselves on the web according to ideas of transparency, usability and accessibility.

'Green' will be treated in the next pages not as a term but as a concept specifically referencing the academic world. We do not intend for this notion to be strictly associated with the fields of architecture or nutrition, as it is in the usual sense. In these two areas, 'green' is connected with the idea of sustainability and environmental protection. We assume that 'green' and 'sustainability' are concepts that can be applied to the academic field in various activities related with student life, including studying, living, doing sports and eating.

B. Producer of The Text: the producer of the texts that are the object of our analysis is a German university which offers an official website where aspects connected to the ideas of green are privileged. The texts belong to the textual genre of online presentations. This genre has key characteristics related to the explicit aim of presenting the institution as the only one the reader should choose, i.e. an optimistic presentation of the institute. Essential features of the macrostructure of the web presentation include:

A) Dividing the content in different sections, with titles as hyperlinks;
B) Describing the best qualities of the university as an educational institution;

4 <www.nielsen.com>

C) Characterizing the university as a factual organization made up of different areas (such as studying, research, teaching, administration) and buildings (e.g. classrooms, libraries, cafeterias, student residences);
D) Offering interactive sections, where interested people can find contacts, social networks (like Facebook, Google+ etc.), contact forms, addresses and maps of the city and of the different buildings;
E) Presenting photography of groups of students, of rooms, of the buildings etc.

The main characteristics of the microstructure of the texts are:

- Use of autosemantic words and expressions with an emotionally positive connotation (e.g. *Offen im Denken*; *Inspiration an Rhein & Ruhr*; *Vielfalt als Potential*[5]);
- Brief sentences that have a paratactic relation. In case of hypotaxis, i.e. sequences of a main clause and subordinate clauses, these are at the first level of subordination in order to have texts with a low grade of complexity (e.g. *Das bedeutet, dass während des gesamten Studiums kein Wechsel in eine gesetzliche Krankenversicherung möglich ist.*[6]);
- Nouns that are the result of a process of composition or of verbal derivation, which are compact and informative[7] (e.g. *Krankenversicherung*, *Bewerbungsphase*, *Studienplatzangebot*, *Studieninteressierte*, *Forschungsort*, *Finanzierung*, *Bewerbung*, *Einschreibung*, *Antrag*).

C. Recipient of The Text: the main recipients of the texts are:

1. Prospective students and their families;
2. Current students who have successfully completed bachelor courses or master courses and who are interested in a higher level of education;
3. People who need to take individual exams or who are interested in attending individual lectures;
4. Professors who work at the university and teachers who work in schools.

The target group of the texts is an audience of prospective experts or of experts, if we consider graduates. In Germany, the autonomous life of university students has a great importance from a social perspective. It is common among prospective students to choose a university located in a city or town away from their family

---

5 Examples taken from the University of Duisburg-Essen, <https://www.uni-due.de/>.
6 Example from the University of Duisburg-Essen, <https://www.uni-due.de/>.
7 See Crestani (2010).

residence[8]. Online presentations of universities help students choose which institution to attend, where they will study and live for the next years of their life. For these reasons, a presentation that places students and their needs at the center will have greater success and prompt more students to apply to the university.

In conclusion, the idea of 'green' applied to communication in the academic context concerns practical aspects of student life such as living, eating and studying. By associating the value of 'green' and the act of 'academic communication,' and by analyzing the connection between the two aspects, we intend to make a timely contribution to the description of institutional communication, which is oriented on the concept of advantages for all people.

2.2 Utopia

The noun *utopia* has a relatively long but extremely interesting history, which is outside the scope of this article. For our purposes, the term itself gives the idea of 'no place', as it is made of two ancient Greek elements with the meaning of negation (οὐ 'not') and of place (τόπος 'place') and the classical Latin suffix *-ia*. It was coined by Thomas More in his *Libellus vere aureus nec minus salutaris quam festivus de optimo reipublicae Statu deque nova Insula Utopia* (1516). In modern dictionaries such as Collins (<www.collinsdicionary.com>) for English and Duden (<www.duden.de>) for German, its definition connotes characteristics of dream, perfection, happiness for everyone and idealist features, which do not belong to real places, and which are typical of an imaginary society. Despite this idea of perfection, Thomas More leaves open the question of whether a perfect society is really ideal.

In the following essay, we explore the concept of utopia in relation to the academic world. In this world, it is the university (with its role as the producer of the presentation) that presents ideals as its fundamental characteristics:

- Fairness in education.
- Equal treatment.
- Equal opportunities.
- Meritocracy.
- Open-mindedness.

---

8 See the article *Erstsemester mit Heimweh: Ausziehen!* (*Der Spiegel*, 17.10.2014): <http://www.spiegel.de/unispiegel/studium/studium-erstsemester-studenten-erzaehlen-ueber-heimweh-a-997128.html>. 40 percent of German students who attend their first semester at the university live with their parents; 60 percent leave their families and live in another city or town.

The presentation also displays characteristics of auto-representation, which references real data for some aspects (e.g. number of students, professors, or apartments in residences etc.) but also abstract concepts (e.g. the challenge of equal opportunities).

2.3 'Green Utopia': A Definition for the Academic World

'Green utopia' is a concept assuming that university students can study and live in a sustainable way for themselves, but also for their current and future society. We use 'green' as the idealistic metaphor for the sustainability-concept and the concept of sustainable development.

Sustainable development "is a development that meets the needs of the present without compromising the ability of future generations to meet their own needs" (1987, World Commission on Environment and Development). Transferring this concept into the world of the university, we describe the educational process as development that connects the needs of the current students with the needs of the (future) society. The needs of the students and of the society include educating students to work and live according to principles of equality, progress and awareness of the needs of the other citizens. University students can undertake the following activities:

- Studying, i.e. acquiring knowledge and developing interdisciplinary abilities and skills, which form the basis for a career;
- Using different areas of the university such as libraries and cafeterias (where they can borrow books, use Wi-Fi to study, eat at the cafeteria and choose a variety of healthy, low-cost meals etc.);
- Taking part in university life;
- Living in a place with a comfortable atmosphere for studying, meeting other students and participating in cultural activities.

3. Object, Terminology and Method of Analysis

3.1 Object

In our paper, we concentrate on the linguistic analysis of the official website of the University of Duisburg-Essen (<https://www.uni-due.de/de/>, last accessed: October 2015). The reasons for using this university website include the following:

- It is a user-friendly site, which should allow students in search of information to find the most important points in an easy way;

- It offers a good overview of the academic world by adopting not only the verbal component but also the eidetic component of the language, which is fundamental in communication with citizens as a heterogeneous group;
- It presents a young institution, whose foundation dates back to 2003 as the result of the fusion of the University of Duisburg and of the University of Essen. The new institution had to adopt cooperation criteria and sustainable criteria to deal with the fusion in the best possible way.

In order to conduct a deep examination of the website, we will provide insight into all sections and we will include not only the online component of the site, which is dynamic and can/must change, but also electronic written documents, which comprise the static part of the site.

3.2 Terminology

We use terminology that may not be familiar enough to untrained non-specialist readers, or which assumes a particular meaning in our article. In order to avoid confusion, we will give a brief definition of the main terms used in our essay.

First, our analysis is linguistically oriented and concentrates on the verbal component of the website, but it also considers the eidetic component by adopting a semiotic interpretation of the web content. The verbal component is in German[9]; the eidetic component includes graphic elements such as colors, photography, characters etc. We consider the relationship between the verbal and eidetic components by using a linguistic-semiotic perspective[10]; this perspective allows us to analyze words not only in relation to internal characteristics (meaning, connotation, word-formation etc.) but also in relation to their external characteristics displayed on the website (color, size, position in the page etc.). The analysis of the connection between the verbal and eidetic components is a good way to understand the principles and ideas expressed on the website.

Second, we consider the website a text. Defining the text from a linguistic point of view may lead to long discussions, which is beyond the purposes of our research. We refer to De Beaugrande and Dressler (1981) and to Brinker / Cölfen / Pappert (2014) for a linguistic definition of 'text'[11]. In our paper, we adopt the concept of 'multimedia text' (see Laskewicz 2004):

---

9 The website offers an English translation or a summary for the most important sections (e.g. its online application and information on living in a student residence).

10 See Brucculeri (2009) for an example of semiotic analysis in the field of tourism.

11 De Beaugrande and Dressler (1981) assume that a text is a "kommunikative Okkurrenz", i.e. a communicative occurrence that meets seven principles of textuality:

- The text is made of a verbal and an eidetic component, which are strictly connected and have a dynamic relation. The eidetic component is the dominant one, as it allows access to other verbal parts through the use of hyperlinks, which have a different color and are underlined. In addition to this informative function, the graphic component has a persuasive and expressive function: through the use of colors, pictures, characters etc., the reader is persuaded to read the text, which expresses the auto-representation of the institution;
- The verbal part is organized as a multilevel component (the different levels are obtained through the eidetic part). Each level (i.e. each hyperlink) represents a deeper level of information and content. It has a core thematic argument (i.e. the specific university) and related arguments, which can be found in separate sections in the website or on other websites, which are explained verbally and visually.

We consider the text as:

1. "Signifikante Praxis" (Barthes 1973), i.e. the activity of giving meaning to the text, when a person reads it or listens / sees it (in case of videos).
2. A dynamic product: it is a medium of communication and of representation. The language in the text is to be understood and to be deconstructed by the reader, who begins to 'play' with the different meanings of the text. That same reader then reconstructs the text.
3. Intertext, i.e. as a transformation of other texts that were given a meaning from the reader and then reconstructed[12].

3.3 Method of Analysis

A text, as explained in paragraph 3.2, needs to be analyzed with a multimodal method in order to avoid considering the text only as a sequence of words,

1. Cohesion; 2. Coherence; 3. Intentionality; 4. Acceptability; 5. Informativity; 6. Situationality; and 7. Intertextuality. If one of these principles is not present, the entity is a non-text and is not communicative.

12 See Barthes (1973 : 1015) : "Tout texte est un intertexte ; d'autres textes sont présent en lui, à de niveaux variables, sous de formes plus ou moins reconnaissables ; les textes de la culture antérieure et ceux de la culture environnante ; tout texte est un tissu nouveau de citations révolues. […] L'intertextualité, condition de tout texte, […] ne se réduit évidemment pas à un problème de sources ou d'influences ; l'intertexte est un champ général de formules anonymes, dont l'origine est rarement repérable, de citations inconscientes ou automatiques […]."

sentences and clauses that have precise semantic, syntactic, textual and pragmatic relations to each other.

Multimodality assumes that communication (realized through written or oral texts) has to be described in terms of:

- A multiplicity of modes, i.e. linguistic, spatial, and visual modes used to create the text;
- The relationships between the modes.

In other words, the approach of multimodality sees communication not only as language, but as the interrelation between different modes, which are more than the single verbal mode. Multimodality is based on two main assumptions:

1. Communication exploits various modes (e.g. the visual, gestural, spoken, and written) depending on different contexts, all of which contribute to the final meaning. The full range of modes has to be analyzed in order to describe how people make meaning.
2. Modes are organized sets of semiotic resources that people use to make meaning. They are socially relevant and culturally shared.

Multimodality is a concept that is often used for the study of oral communication, particularly face-to-face-conversations. We nevertheless assume that multimodality is an effective method for analyzing web-communication. On the web, communication practices are neither simply written or oral; they take on mixed forms and can be written with features of oral communication (e.g. chats, instant messaging) or can be oral with features of written communication (e.g. video-tutorials)[13]. In the case of online presentations (e.g. presentations of universities, of organizations or of companies), there is a visual and often an audiovisual aspect (a video is assumed to be communicative and effective). That is the main reason why we propose multimodality as method of analysis.

4. University Presentation as a Multimodal Object

In the following paragraphs, we present the results of our linguistic-oriented analysis by adopting a thematic criterion. This means that we do not offer an analysis of single sections of the site, but we evaluate the different sections according to three specific themes: living, studying and eating.

13 See Siever, Schlobinski and Runkehl (2005) for communication in chats and Crestani (2015a) for communication in video tutorials.

These topics were chosen due to their relevance for university students: as it was said, a large percentage of students in Germany live in the place where they attend university. Because they spend a good deal of time on campus studying and attending classes, they often tend to eat at the university. For each of the three topics, we consider the linguistic and visual strategies that contribute to ideas of perfection, happiness, and opportunity for all students. We will demonstrate that the website is really only accessible for a group of people. The ideal reader of most sections of the website should be:

- German-speaking or English-speaking: with at least a B2- or a C1-level of proficiency according to the Common European Framework of Reference for Languages.
- Not disabled: blind students cannot access content of some parts of the site and deaf people cannot hear the video presented on the home page, because there are no subtitles or audio-description of the libraries[14].
- Technologically competent and equipped: the reader must have an Internet connection and a PC, a tablet or a smartphone to access the website.

4.1 Living

The term *living* refers primarily to the state of life in a student residence or in an apartment with other fellow students. It also relates to a student's experience of a place by going to the cinemas, theatres, pubs and restaurants etc. and by taking part in the cultural and social life of the city.

4.1.1 Experience the City

For this topic, there is extensive use of emotional words, whose meaning is directly connected with the action of doing something pleasant and worthwhile. Therefore, living in Duisburg and in Essen is an *Überraschung* 'surprise'[15]

14 <https://www.uni-due.de/ub/abisz/audioguide.shtml>. In another section of the library presentation (<http://duepublico.uni-duisburg-essen.de/go/ub-film/Mit_der_Bibliothek_durchs_Studium_1.mp4Z>), there is a video that is primarily visual, as it shows how a student can find good books in the library to do his homework without any audio-description. This video is accessible to deaf people, but not to blind people. Moreover, it has mainly a persuasive function and not an informative one, as it is in the audio-description of the libraries.

15 The symbol "is used for the official English translation offered by the University. Words and expressions that do not have an official translation, will be explained with".

(<https://www.uni-due.de/de/universitaet/leben.php>), because the city has a *reichhaltige Geschichte* 'rich history'. Students can *nachspüren* 'follow' the history of the city when they visit the new *Ruhr-Museum*. Here, there is a fine play on words and concepts between the ancient origins of the city and the city's new and innovative turn in the last decades. Duisburg offers not only history but also a direct link with nature and animals; the zoo is *legendär* 'legendary', and its brief description gives readers the impression of visiting an exotic land with pandas and dolphins, two animals that evoke positive impressions because they are symbols of sustainability[16]. Culturally speaking, Duisburg and Essen are well developed; the *Theater- und Musikkulturszene* 'theater and music scene' offers *europäische Spitzenklasse*, i.e. it is excellent and of European standard. By living in these cities, students are provided with good entertainment opportunities such as a *lebendige Kneipenkultur* 'lively pub culture' and *niveauvolles Shoppen und Genießen*, i.e. stylish shopping areas and culinary highlights.

The takeaway message is that Duisburg and Essen are very nice places to live because they offer many experiences to live life to the fullest. The producer of the site is aware of the emotionally charged experiences of culture, nature, shopping and going out with friends, and he / she uses these experiences to create a text that proposes a sustainable way of life.

The history of the region is ancient and today it is a lively, thriving area. Up to the 1970s, however, the Ruhr metropolis was perceived as a dirty, cold region, based on the coal and steel industries since the nineteenth century[17]. The image of a grey city, damaged by pollution and exhaust fumes, is replaced with the image of green areas in the city and around the city, with open-air cafés packed with people enjoying life (see <https://www.uni-due.de/de/universitaet/leben.php>). The sky, which can be seen in the photography featured on the site, is blue with white clouds, a positive image testament to the environmental protection work that has been successfully undertaken since the 1970s. White clouds are in

16 See the logo of WWF with the giant panda as symbol of the environment; the dolphin was used in the Greek mythology with a positive connotation.

17 In the 1960s, air pollution was a visible problem in the Ruhr area due to the dust, ash and soot from furnaces, steel converters and coke ovens. Willy Brandt stated in the speech at the Bonn Beethoven Hall (April 1961) that "der Himmel über dem Ruhrgebiet muß wieder blau werden" / "the sky over the Ruhr region must be blue again." This statement represents the start of systematic environmental thinking in German politics. Since the 1970s, this challenge has been met with success. See <http://www.umweltbundesamt.de/en/press/pressinformation/federal-environment-agency-sky-over-ruhr-is-blue> for more information on pollution today.

antonymic relation with the grey fumes of the industries. The color of the most important part of the site is also blue, which for a contemporary audience connotes elegance, simplicity, refinement or something official. According to Agnello (2013: 26), the blue color is "il colore preferito dalle istituzioni (ONU, Consiglio d'Europa, UNESCO ecc.) nonché il colore ufficiale delle aziende che vogliono veicolare serietà e tradizione [...]." Institutions as well as companies often use blue to communicate values of seriousness and tradition. We can assume that the reader will associate the emotionally connoted words with the sense of something official, of an objective description of the city and of a reality that is described. That motivating force leads to a utopia. The reader is involved in a play of words and mental associations, which encourage him / her to think about Essen and Duisburg as utopian microcosms, where he / she can live in harmony with himself / herself, with other people and with the environment, keeping in mind the affluence of the region is due to its industrial history. This is the weak point of the utopia hidden behind the text: the photography represents a green world while the words refer to the region's economic power. People cannot forget that the region is based on heavy industry, although good results have been achieved in protecting the environment. Despite their awareness of this region's industrial legacy, recipients can substitute the idea of a drab, grey place with the idea of happiness, because the positive graphic component of the presentation is dominant and they are strongly influenced by the power of the visual communication.

The idea of happiness and of the best opportunities that students can find in the region is the thread of all related topics of the experience.

4.1.2 Experience the University

The university offers opportunities in sports, music, theater and cinema. Their description is based on expressions like *ein breites kulturelles Angebot* 'a broad cultural offer' (<https://www.uni-due.de/de/studium/kultur_div.php>), where the adjective *breit* 'broad' refers to the idea of an educational concept of European openness. In the section, sports in nature such as sailing and boating competitions in which both students and professors get involved are emphasized:

> 1. Beim gestrigen Essener Firmenlauf waren insgesamt 200 Studierende, ProfessorInnen und MitarbeiterInnen aller Bereiche unserer Universität angemeldet. Wieder gelang ein hervorragendes Ergebnis. "In yesterday's race in Essen, there were 200 students, professors and employees from all areas of our University. Once again we achieved an excellent result."

The idea of a very good result (*ein hervorragendes Ergebnis*) is related to the gender-orientated use of the plural: *Studierende*, *ProfessorInnen* and *MitarbeiterInnen* include both male and female people. This plural form (instead of forms like *Studenten*, *Professoren* and *Mitarbeiter*, which refer only to men) is intentionally used to avoid gender discrimination[18]. The field of *Genderlinguistik* (gender linguistics) is a sociolinguistic field of analysis that was developed in the 1970s with the aim of finding out whether language differences between women and men can be explained through the different roles they have in the society. Furthermore, this field attempts to develop rules to avoid sexist use of the language. The category of natural gender plays an important role in our society and in the language used, as observed by Spieß, Günther and Hüpper (2012: 1): "Das 'altertümliche Dual' prägt weiter unsere Sprache, unsere Mode und nicht zuletzt die Gehaltsverteilung in unserem Land." According to the three authors, the medieval dual male-female continuously influences our language, our practices and – last but not least – the differences in salary in Germany.

In the sport section of the university website, photography is the most meaningful part, with a predominance of blue and green colors (including the sport uniforms of the group of students and professors taking part in a competition). Blue and green refer to open-air sports.

In addition, the section about the *Essener Studentenorchester* (<http://www.eso.uni-due.de/>) underlines values like liveliness and cooperation among (former) students: *Wir sind eine lebendige Truppe aus aktiven und ehemaligen Studierenden* "We are a lively group of students and ex-students". Here the word *Studierend* instead of *Student* ensures that both male and female students are taken into consideration. Another important value highlighted from a verbal and a visual perspective is the attitude of openness towards other students, who are invited to join the group of musicians. The concept of invitation is repeated three times with the following visual organization. First, it is integrated in the text and second, it has a special section, a box with a yellow background, and it is written in the same font (in bold). Finally, the verb *mitspielen!* "play in the orchestra with us!" is written in a red color (red is used to attract the user's attention). This word is separate from the text and it is marked in italics. The italics extend a personal invitation to the reader, as the curvy font conjures something hand written, perhaps even by the conductor of the orchestra. The meaning is "You are invited to play with us".

This element involves the reader, and given the presence of the button *Gefällt mir* "like" for Facebook at the end of the section, the reader can actually respond to the content.

---

18 See Miemietz (1993), Schwarz (1999) and Robustelli (2000). Miemietz and Schwarz in particular deal with possibilities and borderlines of the morphological motion in the person designation and in the job designation.

The invitation to be part of university life is present in other sections such as the *Universitätschor Essen* (<https://www.stud.uniessen.de/~sh1053/>), where the invitation is open to students who fulfil the following three requirements:

Singst du seit deiner Kindheit und immer wieder mal Stücke von Musicals?

Gerne verwandelst du dich beim Singen in einen Star wie Rihanna oder Ed Sheeran?

Dann sind auch alle Texte der Wise Guys eine Leichtigkeit für Dich?!

These three questions refer to a growing level of skill, but most of all they act emotionally on the reader and invoke their dreams (e.g. a childhood dream of singing, the dream of being a famous singer like Rihanna or Ed Sheeran, or the dream of singing difficult lyrics like those of Wise Guys): "Have you been singing songs from musicals since your childhood?"; "Would you like to use your singing transform yourself into a star like Rihanna or Ed Sheeran?"; "So all songs by Wise Guys are easy for you to sing?!". The three questions are also graphically depicted as growing in difficulty; the increase of difficulty is underlined by marking the first words of the questions in bold, i.e. *Singst du* "Do you sing", *gerne* "like" and *dann* "so". The last question is intended as a direct consequence of the positive answers to the first two questions (if you've been singing since the childhood and you want to become a star like Rihanna or Ed Sheeran, then you are probably able to sing songs by Wise Guys).

In the English theater section (<https://www.uni-due.de/duet/>), the sense of invitation to be part of the university is less clear from a verbal perspective. The text always has a positive connotation in the visual component: the background is blue with tones of yellow and green.

Green and blue are colors that reference sustainable development at the university (<https://www.uni-due.de/ifn/>). Students are invited to take an active part at the *Initiative für Nachhaltigkeit – Universität Duisburg – Essen*, which has three important keywords, all the results of word-formation-processes (composition and derivation), and all convey a high degree of information with syntactic compactness:

- *Zukunftsfähigkeit* 'future potential'
- *Lebensqualität* 'quality of life'
- *Nachhaltigkeit* 'sustainability'

In this initiative, students can actively participate in events and discuss the preservation of the environment with experts to guarantee a good quality of life for both the present and future generations. This project is described using positive words and expressions, which has two main functions:

– A persuasive function, such as *wir laden Euch herzlich zu unserer Veranstaltung „Corporate Social Responsibility im Dialog“ ein* / “We warmly invite you to our event “*Corporate Social Responsibility in dialogue*””, where the reader is directly invited to the event.

– A connotative function, such as *Liebe Freunde der Initiative*, where the reader is considered a friend deserving of a personal invitation addressed to the recipient of an informal letter or newsletter (“Dear friends of the initiative”).

In this description of the initiative, the utopian background of the project becomes evident. The reader is defined as a friend who can cooperate with other friends to make suggestions about sustainability and to discuss important topics. The reader is thus involved in a utopian world created by the producer of the text. From November 2014 to October 2015, the text was not changed[19]. A particularly careful reader would conclude that the initiative did not have a great deal of success, and that concrete actions to deal with an important topic like that of sustainability are probably taken only once a year. The symbol for the initiative is also utopian: a circle with light green contours and interconnected lines, which represents the earth and its connections to different places. The idea of connection among people is really a utopia; however, the only possible way for the reader to get in touch with the people organizing this initiative is with an email-address, which is difficult to find, as it has been placed at the end of the page and is written in a small font.

The concept of sustainability also assumes utopian contours in other fields, such as living in a dorm and eating in the canteen, as will be explained in the next paragraphs.

4.1.3 Living in the City

Students who must leave their family home search for a place to live that is comfortable, perhaps with other students, and where the costs for rent are not very high. The website of the University of Duisburg-Essen offers information about and descriptions of the student residences that belong to the university. The *Studentenwerk*, a sort of student union that provides financial, social and medical services for students, runs the student residences (<http://studentenwerk.essen-duisburg.de/wohnen/>). The student residences present general information on cost of rent, number of rooms, types of rooms and the furniture provided. Expressions like *saniert* ‘updated’ and *modernste Standards* ‘the most modern standards’

19 The text was then modified in October 2015 to present a new event on the topic of the *Zukunftsstadt* “city of the future”.

often appear, which reference energy efficiency requirements. They are, however, the only references to the idea of sustainability, which seems to be important for the institution (as explained, a single section of the web site is dedicated to the *Initiative für Nachhaltigkeit – Universität Duisburg – Essen*). In some cases of the student residence descriptions, there are also more technical words borrowed from the language of bioarchitecture. Some examples include *wärme- und geräuschdämmend* 'thermal and noise insulation' and *Photovoltaikanlage* 'photovoltaic system'.

These compound expressions all refer to important aspects of sustainable buildings, such as the thermal and acoustic characteristics. The theme of sustainability concerns not only the material aspects of buildings, but also accessibility for disabled people and foreign students. This is, however, also utopian: the reality speaks against the utopia of sustainable opportunities for these people. Only one residence has an apartment that is wheelchair accessible. Foreign students also have difficulties finding a room, because they tend to have fewer financial resources in comparison to German students[20]. Accordingly, a prospective student reading this website would get a different impression about the university's capacity to welcome students in need of special accomodation. German students themselves can also have difficulties finding a room, as there is a gap between the number of students and the number of available rooms financed by the state:

> 2. „Die Schere zwischen der Zahl der Studierenden und der Zahl der staatlich geförderten Wohnheimplätze geht immer weiter auseinander", erklärt DSW-Generalsekretär Achim Meyer."[21] "The gap between the number of students and the number of rooms financed by the state is becoming increasingly bigger", as the general secretary of DSW, Achim Meyer, explains."

On the homepage of the university, there are expressions like *offen im Denken* 'open-minded' (which is the motto of the institution) and *Vielfalt als Potential* 'Diversity – Realizing potential'. They all refer to values like:

- Open-mindedness: this concept has many semantic interpretations and it can be understood as an umbrella-concept for a wide spectrum of ideas (e.g. fair education, good educational opportunities for all people, combating discrimination and homophobia). As an abstract concept with many related

20 See the article „Wohnungssuche ist das größte Problem ausländischer Studierender in Deutschland" (22.07.2015), <http://studentenwerk.essen-duisburg.de/nc/wohnen/newsdetails/archive/2015/juli/artikel/wohnungssuche-ist-das-groesste-problem-auslaendischer-studierender-in-deutschland/>.

21 <http://studentenwerk.essen-duisburg.de/nc/wohnen/newsdetails/archive/2015/oktober/artikel/mehr-preisguenstiger-staatlich-gefoerderter-wohnraum-fuer-studierende/>

concepts, open-mindedness represents the prototypical example of a utopian concept.

- Diversity as potential: the goal of achieving diversity in society is a widespread idea. Diversity can refer to:

A. Gender equality among men and women: good examples are linguistically inclusive-forms (e.g. *Studierende*), which include both groups.

B. Inclusion of disabled people: the most important word is here *Inklusion* "inclusion", a derived word with a specialized meaning in the field of the educational sciences. It is an approach in which students with special educational needs spend most of their time with non-disabled students (see Baglieri and Shapiro 2012). The deverbal noun *Beratung* "advisory service" is a keyword here:

> 3. Schwerpunkte der Beratung sind in der Regel Nachteilsausgleiche im Studium, wie im Bereich der Studien- und Prüfungsmodifikation, der Hochschulzulassung […].[22] "Important issues for the advisory service are usually balancing out disadvantages in the course of study, e.g. modifications to examinations, to access to the university […]."

The concept of *Barrierefreiheit* 'accessibility' is also very important. The word is a compound of 'barrier' and 'freedom,' referring to accessible rooms at the university, but also to the possibility of finding an adequate room or apartment to live in (<https://www.uni-due.de/inklusionsportal/wohnen.shtml>). Disabled people are not only *rollstuhlfahrende und mobilitätsbehinderte Studierende* "students in wheelchairs and students with mobility problems", but also *sinnesbehinderte und chronisch kranke Studierende* "sensory impaired and chronically ill students". The two groups of students are described as people with the possibility of *eine bevorzugte Berücksichtigung* "preferential treatment". Students in need of a wheelchair or blind students need to be able to enter a classroom and have accessible solutions for ICT products. There is a sequence of words related to the concept of *Barrierefreiheit*: *barrierefreie PCs* "barrier-free PCs", *barrierefreie Notebooks* "barrier-free notebooks", *barrierefreie Software* "barrier-free software", *barrierefreie Plätze* "accessible seats", *digitale Barrierefreiheit* "digital accessibility".

C. Integration of foreign people: Particular attention is paid to refugees (<https://www.uni-due.de/international/wege_an_die_ude.php>) as consequence of the dramatic influx of new refugees from Syria and other countries in recent years. The policy of Germany has been to welcome them and to offer them the chance to study. The words *Weiterbildung* 'advanced training', *Nachqualifizierung* 'qualification' and *sportliche Aktivitäten* 'sports' are used to explain to refugees what they

---

22 <https://www.uni-due.de/inklusionsportal/beratungsstelle.html>

can find at the university. The website offers different situations for refugees, e.g. if they are *Schüler/in* 'pupils', if they have *schulpflichtige Kinder* 'children are required to attend school', if they were students in their country etc.

D. Equality of homosexuals: here the potential of diversity refers to the concept of equality.

E. Socioeconomic diversity: merit and ability should determine whether a student attends school, not their family's financial background. In addition to outlining available scholarships, the university also offers a video about the *UDE-Stipendienprogramm* "scholarship program of the UDE" (<https://www.uni-due.de/de/marketing/deutschlandstipendium.php>). The video begins with the following sentence:

Früher förderte das Ruhrgebiet Steinkohle.

Heute Talente.

This can be translated as follows: "Once the Ruhr area produced hard coal. Today it produces talented people". The visual background of these words is a blue sky with white clouds and rays of sunlight. This image evokes the idea of sustainable education, open to talented students of all social backgrounds. This idea is represented in the words of the people depicted in the video, e.g. female students receiving the scholarship. One of these women explains that the grant provides very important financial support for students *mit Migrationshintergrund* "with a migrant background", once again emphasizing the utopia of the possibility for migrant students to achieve a very good education at this university. The same student explains that the scholarship represents *eine ideelle Förderung* "ideal support", using words with positive connotations. If we look at the data, we understand that the grant is an important form of financial support, even though it is not enough to cover all the costs for a university student. The video also goes on to explain that 1,422 scholarships were awarded in 5 years, which is good, but not enough to help all talented students. The utopia of great opportunities for all talented students is also recognized in the video, and it ends with the following sentence:

Wir haben noch mehr Talente!

This can mean: "We have more talented students!" or "We also have other talents!". The university's idea is to give scholarships to larger numbers of students or to underline many other positive aspects of its organization.

The list is longer than the examples mentioned above and includes a wide variety of situations.

The diversity portal of the university (<https://www.uni-due.de/diversity/>) contains a specific overview of what the institution understands as belonging under the term *diversity*:

- *Gender* "gender";
- *Familie* "family";
- *Internationales – Interkulturalität* "interculturality";
- *Gesundheitliche Beeinträchtigung – Behinderung* "health impairment – disability";
- *Lebenslanges Lernen – wissenschaftliche Weiterbildung* "life-long learning" ;
- *Bildungsgerechtigkeit* "educational justice';
- *Bildungsaufstieg* "educational success".

In order to express the variety of concepts included in diversity, not only from a verbal but also from a visual perspective, the institution uses an image of a group of stick figures (men and women) in different colors. The reader is first involved in a play of visual decodification, which is then supported by the verbal component of the page. The different colors[23] build a semiotic connection to the idea of diversity. It is surprising, however, that the idea of harmony and friendship among different people, which is well described from a verbal perspective, is destabilized by the image itself. Some stick figures are indeed represented as couples who speak together in a friendly way, but instead of using different colors for the two people involved in the conversation, the same color is used for the single pairs. So one couple is blue, the other couple is red, etc. The reader might understand the image as a contradiction of the verbal text, and as a result consider the values expressed in the text as a sort of utopia. Alternatively, he / she can interpret the figure as a provocation, directly enacted by the producer of the text, who is aware of the gap between the concept expressed in words and the real world.

The presence of negative words like *Hürden* "obstacles" supports this second interpretation of the image described above. These words are used in the same page to describe the project *Essen 2030 Ich mach' mit!*, which aims to make Essen an attractive and future-looking city. The idea of diversity seems to be a fundamental one, as the reader sees by reading the last section of the page, where there is a list of the award-winners of diversity prizes. Here, the keyword of the text is *Diversity-Kultur* 'diversity culture', which is a hybrid compound made of the English word

23 It is interesting that black and grey are not used because of their semiotic meanings, referred respectively to death and moral rigidity (black) and old age (grey) (see Agnello 2013: 38–39, 41).

diversity and the German word *Kultur*. Again, words speak in favor of the positive effects of the diversity and of its importance, as the compound is effectively a union of two languages and two cultures. In another section on the website, the concept of diversity is again emphasized: *Vielfalt als Potential ist ein gelebter Leitgedanke der Universität Duisburg-Essen* 'The potential of diversity is a central idea lived by the University of Duisburg-Essen', referencing the third diversity day in June 2013. Core concepts referred to by the statement are *Chancen* "opportunities", *vielfältig* "diverse", *Austausch* "exchange", *Herausforderungen* "challenges", *Projekte* "projects". The university defines itself as a *Universität der Potentiale* 'university of potential', referencing its diversity management. The poster explaining this concept (<https://www.uni-due.de/diversity/projektposter_audit.shtml>) contains a definition of diversity for the university. This definition is based on:

A. Two-word or three-word compounds with a positive connotation (*Chancengerechtigkeit* 'equal opportunities', *Potenzialentfaltung* 'development of potential', *Diversity-Kompetenzentwicklung* 'competences development according to diversity-principles').

B. Derived lexemes, which also share a positive connotation (e.g. *Innovationen in Lehre und Forschung* 'innovation in teaching and in research', *Öffnung der Hochschule für neue Zielgruppen* 'university for new target groups', *Kooperationen und Vernetzung* 'cooperation and networking'). These words, which are associated with abstract concepts, are supported by more concrete expressions such as *Übersetzung von Formularen des Studierendensekretariats ins Englische* 'English-translation of the student forms', which are a minority part in comparison to abstract expressions. The reader of this text must have a good deal of knowledge about specific concepts (e.g. diversity management) to understand the presentation.

4.2 Studying

The term *studying* refers to following aspects in the presentation under analysis:

- A temporal aspect: the activity of studying is intended as a future action (for prospective students), as a present action (for students currently enrolled at the university) and as a past action (for graduates). Utopian ideas have a particular influence on the group of people who are in search of the best university to study at.
- An ethical aspect: as an institution, the university in general is connected to the idea of meritocracy. All people who (can) achieve good grades theoretically have the right to attend. This universal concept of "all people" is related to hypothetical situations: foreign students must have the same opportunities as Germans; female students must have the same chances as male students to earn

scientific degrees in fields such as mathematics, chemistry, and physics, which are typically dominated by men; and students with few financial resources must be given the concrete opportunity to attend.

- A social aspect: University is seen as a unique institution where people can acquire the knowledge, skills and abilities to critically analyze real events, to find viable solutions to problems and use these competences in their future working life. From a social perspective, knowledge, skills and abilities are fundamental, because graduates can contribute to developing society in a good way, if they use their theoretical and practical background for the betterment of the world (i.e. protecting the environment for current and future society).

These three aspects are interrelated and all contribute to create the utopia of "Studying at the University is a right". The University of Duisburg-Essen is actually an institution that presents itself as open and democratic, with the aim of guaranteeing the opportunity for everyone to study.

Referring to the ethical aspect, the university promotes equality initiatives, such as events inviting women to study engineering, mathematics and similar courses (<https://www.uni-due.de/abz/mmm/>). This initiative has a loaded name: *MINT-MIX für Mädchen* "MINT-MIX for girls", where MINT is an acronym for *Mathe, Ingenieurwissenschaften, Naturwissenschaften, Technik*. This name is written with three different colors: cyclamen, light blue and light green. Light blue and green are the most commonly used color in the website, and these colors represent the idea of sustainability in the field of academic studies. Cyclamen identifies the target group: girls. The distribution of the color has connotative meanings:

- Cyclamen is used for the words MINT and MIX and for the initial letter for the acronym MINT in order to emphasize the integration of female students in courses pre-dominantly attended by male students.
- The other letters in the explanation of MINT are in light blue. The use of color in MATHE, TECHNIK forms a theoretical union of men and women who attend the same courses. The long form of the acronym highlights the first letter in cyclamen to make the first letter more visible to the user. Another possible interpretation of this graphic explanation is that most of the letters are light blue, because most of the students are male (traditionally light blue is coded with male, while pink connotes femininity). According to this second interpretation, the idea of equal opportunities for both genders comprises a utopia.
- The phrase *Für Mädchen* is in light green. The idea of sustainability expressed here is that girls can achieve the same excellence in academic study as their fellow male students, and they can become important professionals in these fields. Light green is also the symbol of female, which is between the words

MINT and MIX. There is a semiotic connection between the symbol and the expression *Für Mädchen.*

'Equal opportunities between girls and boys' is repeated several times on the web site[24]. The MINT-initiative is a project of the *GeCKo*, i.e. the *Gender & Co. Koordinierungsstelle*, a department tasked with coordinating gender initiatives. The main linguistic features used to describe the purposes of GeCKo include:

- Use of rhetorical questions and of declarative sentences, which reference stereotypes in a humorous way (e.g. *Jungs als Grundschullehrer – das passt doch gar nicht! Technik ist doch nichts für Mädchen!? Stimmt doch gar nicht!* "Men as primary school teachers – it's not true! Technology is surely not for girls!? It's absolutely not true!"; *Dass diese Aussage schon lange ein "alter Hut" ist, können Mädchen auch in diesem Jahr wieder hautnah an der UDE feststellen.* "Also this year girls can see here at the university that this statement has been nothing more than a cliché for a long time now".
- Use of complex words such as compounds related to the gender equality (e.g. *gendergerechte Studienwahl* "gender-sensitive course-choice").
- Use of the informal personal pronoun *du* "you" to create a context of personal and direct interaction (e.g. *Hier kannst du schnell und unkompliziert herausfinden, welche Projekte dich darin unterstützen können.* "Here you can quickly and easily discover which projects can support you").

The use of the pronoun *du* is a clear sign of a utopian world, which often only exists on the web[25]. In the reality, all official documents are written with the formal *Sie* pronoun. For example, students who want to apply online read sentences such as *Eine Bewerbung ist erst möglich, wenn Sie Ihr Abiturzeugnis bereits erhalten haben.* "You can apply only after you have received your Abitur certificate" or *Um keine Termine zu verpassen und alle Abläufe im Blick zu haben, drucken Sie sich die Checkliste aus.* "In order not to miss important deadlines and have all steps in mind, print out the checklist." In addition, the verbal tense is different. In the introductory sections to the application, there are imperative forms and verbal

---

24 Additional examples can be found in the section *Bildung: gleiche Chancen für alle* 'Education: Equal opportunities for everyone', <https://www.uni-due.de/de/universitaet/chancengleichheit.php>. In this text, the isotopic sequence is evident in expressions like *eine gerechte Bildung* 'fair education', *jedes engagierte Talent* 'every motivated talent', *zu einem erfolgreichen Abschluss* 'to a successful degree', *unabhängig von Geschlecht, ethnischer oder sozialer Herkunft* 'regardless of gender, ethnic, or social background'.

25 Students would prefer the use of *du*-forms also in the real academic world in the communication between professors and students (Brambilla 2012).

expressions with a deontic meaning (e.g. *müssen* "must", *sollen* "must", *sein + zu +* infinitive "be + to be + participle"). The idea of possibility, which is expressed in the section of gender equality through the modal verb *können* 'can' is substituted by the idea of necessity; the idea of necessity must be respected in order to be admitted at the university.

The opportunity for foreign students to apply to the university also belongs to the ethical aspect. Here it is important to appreciate the fact that information is given in both German and English, which caters to an international audience. The reader can also choose between the two languages and read very important details about applying while still living in their home country. Among the information provided, there is advice about financing the costs of studying and living, which is also of interest for German students on a budget. Financing possibilities for German and foreign students are different, but in both cases, there are some aspects emphasized in funding schemes:

- High performance of students.
- Great talent of students.
- Migrant background.

The aim is to support students to complete their studies successfully, and to promote good students by taking into consideration the different backgrounds and needs of the students. According to this view, the student as a person with his needs, problems, abilities etc. is the center of the academic communication. The communication is here very effective, as it describes a world in which strong performances and talent are taken into great consideration. Accordingly, the reader understands that excellent results (together with a difficult background) are the keys to obtaining a scholarship and/or financial support. Again, a utopian world is described here:

> 4. Mit dem Stipendium werden in diesem Förderzeitraum 336 herausragende Studierende für ihre Leistungen belohnt. Insbesondere leistungsstarken Bildungsaufsteigerinnen und Bildungsaufsteigern wird so eine realistische Chance auf ein erfolgreiches Studium geboten.[26] "The grant is for 336 excellent students, who will be rewarded for their performance. In particular, the grant offers educationally ambitious students with the best performance to complete their studies with success."

The words used here have a positive connotation related to the semantic field of excellence and to the real chance to be successful: *herausragend* "excellent", *leistungsstark* "with good performance", *realistisch* "realistic", and *erfolgreich*

---

26 <https://www.uni-due.de/de/marketing/deutschlandstipendium.php>

"successful". The number of students who can obtain a scholarship, however, is low when compared to the number of students who attend the university. Indeed, of the 41,160 students, only 336 receive scholarships – a very low percentage. In this section, which should be realistic and based on real data, there is a significant use of optimistic words and expressions:

5. Wir denken in Möglichkeiten statt in Grenzen. 'We have a fresh way of looking at things.'

6. Wir sind stark in Forschung und Lehre, leben Vielfalt, fördern Potentiale […]. 'We are strong in research and teaching, embrace diversity, promote academic potential […].'

In the examples, there are antonymic expressions (*Möglichkeiten* "possibilities" vs. *Grenzen* "limits") and the university identifies itself with the positive expression (i.e. "possibilities") and it builds a sequence of positive expressions in the text. This semantic isotopy[27] gives the institution the image of a meritocratic institute, where the abilities of people and not their social or geographical background are taken into consideration. In another section, the idea of educational sustainability is emphasized:

7. Es werden junge Talente aufgebaut, die Region gestärkt und gesellschaftliche Verantwortung übernommen. "Young, talented people are supported, the region is developed and social responsibility is taken."

By reading further sections on the website, the reader is informed that the amount of a grant given by the University is 300€ per month (<https://www.uni-due.de/de/marketing/deutschlandstipendium.php>). Although the amount is not low, it is not enough for a student to cover all the costs of studying, particularly when we consider that a room in a residence hall costs at least 230€ per month. A student who cannot receive financial support from his / her family must therefore find other financial support. Equal opportunities are simply not guaranteed to all. The university describes itself as one of the most successful German universities in the field of scholarships. This is an objective statement, based on quantitative comparisons and rankings. The utopia is thus present, as explained above. The idealization of the scholarships as a guarantee for talented students is also emphasized in the photography presented in the specific section for the scholarship application (<https://www.uni-due.de/de/marketing/studierende.php>). The photography has the communicative function of introducing the text, where students can find out about the required information and the prerequisites for the application. In the photograph, there are three students (from left to right: boy,

27 See Heinemann (2000).

girl, boy) who stand in a line holding a tray in their hands with a fork, a spoon, a knife and a paper napkin on top of a place mat. The figure on the place mat has an associative relation with the photography present in all sections of the website: a blue sky with white clouds. It is evident that semantic isotopy is not only produced in the verbal component of the website but also in the graphic component. The relation between photography and text is one of denotation. The photography can be interpreted in two ways:

A. Students stand in a line waiting for the results of their application; and B. The three students have received their scholarships. They now have a specific position in the list and are metaphorically waiting to attend lecturers and learn.

In both interpretations, it is evident that only a few students can receive the scholarship: there is a ranking and not all students with good marks will obtain the scholarship. The importance of the photography for the text is visible in the corresponding section in English (<https://www.uni-due.de/de/marketing/students.php>). The text is here less informative, as it is semantically reduced, and the photography is different from that in the German section, but it is still extremely communicative. Here, five students (four girls and one boy) are together walking outside, but only one of the students is of a clear non-German ethnic background. The idea of educational sustainability, as expressed in the German version, is abandoned here.

4.3 Eating

Nutrition is a field in which sustainability is a value of great importance (see Crestani, accepted). The presentation of canteens and cafeterias (<https://www.uni-due.de/de/services/mensen.php>; <http://studentenwerk.essen-duisburg.de/gastronomie/>) is based on schematic descriptions of opening hours, menus and costs. A short description of the buildings where the canteens are located is also offered. The concept of sustainable food is highlighted in the general presentation of the canteens offered by the *Studentenwerk*:

> 8. Wo dies möglich ist, setzen wir auf ein regionales, saisonales und biologisches Angebot. So verarbeiten wir ausschließlich Bio-Nudeln und Bio-Reis und beziehen ausschließlich Fair-Trade Kaffee, in der Regel auch in Bio-Qualität. [28] "When possible, we serve regional, seasonal and organic food. We use organic pasta and rice and we buy exclusively fair-trade coffee, which is usually organic quality."

28 <http://studentenwerk.essen-duisburg.de/gastronomie/>.

In all canteens, attention is payed to products like coffee, pasta and rice, which are certified as organic. The reader can immediately recognize which products on the menu are certified with the logo.

The idea of green is also really emphasized in the cafeterias, for example in the cafeteria *ins Grüne* (<http://studentenwerk.essen-duisburg.de/gastronomie/mensen/bistro-insgruene/>):

> 9. „insgrüne"- die Duisburger Mensa hält, was der Name verspricht. Die Räume an der Bismarkstraße präsentieren sich lichtdurchflutet und stilvoll im Grundton Grün eingerichtet.[29] "'insgrüne' – the canteen in Duisburg holds the promises of its name. The rooms in the Bismarkstraße are flooded with light and are rich in style with their light green color."

The general presentation of cafeterias underlines sustainable aspects through keywords and expressions such as *Bio-Kaffee* "organic coffee", *aus fairem Handel* "fair-trade goods", *die Schonung von Natur und Umwelt* "the conservation of nature and the environment", *eine sorgfältige Müllvermeidung bzw. -verminderung* "proper waste prevention and reduction", *Mehrwegflaschen* "returnable beverage bottles". Compounds in the field of sustainable food often use the confix *bio-* as first element or deverbal nouns as second elements to build *Rektionskomposita* (synthetic compounds, see Gaeta 2010 and Crestani 2015b). These compounds are informative and very compact; they substitute entire clauses (e.g. *Müllvermeidung* → Es wird vermieden, Müll zu produzieren) and they help to describe the ideal of sustainable nutrition. The *Studentenwerk* presents itself as a sustainable institution:

> 10. Das Studierendenwerk Essen Duisburg ist biozertifiziert durch die DE-ÖKO-039-Kontrollstelle. Der Einsatz fair gehandelter Kaffeeprodukte sowie die Berücksichtigung regionaler Anbieter sind für uns selbstverständlich.[30] "The *Studentenwerk Essen Duisburg* is organically certified by the DE-ÖKO-039-Kontrollstelle. The use of organically produced fair-trade coffee-goods and the consideration of regional farmers are our principles."

The idea of regional products is one of the principles of sustainable nutrition (see Crestani, accepted) and it is frequently highlighted in the website (e.g. *regionale Lieferanten* "regional suppliers", *regionale Produzenten* "regional producers"). Other principles include the conservation of the environment (*Umwelt- und Ressourcenschutz*) and recycling methods and recycled products (*Recyclingpapier, Mehrwegflaschen-System*).

Despite paying great attention to the processes, nutrition, in the sense of what food to eat, receives little attention on the website. The lack of information about

---

29 <http://studentenwerk.essen-duisburg.de/gastronomie/mensen/bistro-insgruene/>

30 <http://studentenwerk.essen-duisburg.de/gastronomie/kundeninformationen/umwelt-und-ressourcenschutz>

nutrition is utopian in a negative sense because students need to be informed about the best food for their health and for the preservation of the planet. These days, a large number of students are interested in a healthy diet[31]. It also seems ineffective that an informative section about diet is not included in the university presentation.

5. Conclusions

In our paper, we discussed how the concepts of 'utopia', 'sustainability' and 'education' can be analyzed together in the specific communicational context of online presentations. The hypothesis that guided this study was that online presentations of institutions, which are important for specific groups of citizens such as young adults and their families in the case of university websites, are a very good place where utopic ideas and concepts can be expressed. To verify our idea, we concentrated on the website of the University of Duisburg-Essen, a very young institution of higher learning, which had to find effective modes of communication in order to reach a wide audience of users interested in becoming students.

In order to analyze how the institution communicates utopian ideals in their online presentation, we adopted the method of multimodality, using the following principles to orient our study:

- The online presentation is a multimodal object that can be studied as the interrelation of several modes of communication (visual and verbal, visual and non-verbal, visual and acoustic etc.) and of several levels of construction.
- The user of the website is the subject who interprets the meaning of the text using a decodification system in order to discern meaning and give an interpretation of the modes of communication and the levels of construction of the text.
- The act of making meaning is an act that depends on cultural values and on the social conventions present in the text, as well as the values adopted by the user.
- We offered an analysis that considered the different modes of communication and their interrelations. The interrelations can be summarized as follows:

31 See <http://www.asu-arbeitsmedizin.com/ASU-2008-11/Studie-zur-Ernaehrung-von-Studentinnen-BRTeil-II-Wie-ernaehren-sich-Studentinnen-in-Deutschland,QUlEPTIyNDA3OCZNSUQ9MzAwMTA.html>. According to a survey conducted among female students in Germany in 2007 about their nutrition habits, the students periodically consume organic products; more than 50% consume fruit, salad and vegetables at least once a day and only few eat fast food meals. Many students (in particular those who are overweight) often consume sweets and snacks. Despite this, the conclusion of the survey is that female students are generally interested in a healthy way of eating.

- Symmetric: the modes of communication are interdependent and they are in a relationship of support with each other. An example is the text on the MINT-MIX project, where the graphic and the verbal parts follow the same principles, with the graphic component emphasizing the verbal component.
- Antonymic: the modes of communication are in opposition. An example is the opportunity for foreign students to apply for a scholarship from the university. They are allowed to apply (as explained in the rules), but the photography contradicts the possible results of the application (there is only one foreign student in the group of five students).
- Scalar: the modes of communications are used to express meanings, which are in increasing degrees of a range. An example is the project of fair education. This value is expressed on the university's page of information for prospective students and it is increasingly emphasized in the different sections and subsections of the website, in both verbal and non-verbal aspects.

The utopian ideals are visible in particular when the modes of communication are in antonymic relation. Utopian communication is also realized with the symmetrical and scalar relations; in order to understand the hidden utopia, the user has to carefully read all sections of the presentation (including external links) to decode the text as a whole.

In conclusion, when it comes to communicating about fairness and sustainability in education, positive words and expressions can be effectively used to create a utopian world. This world exists in the verbal component of the presentation, but it is sometimes 'destroyed' in the graphic component of the website. Two examples of this include:

- The verbal component that speaks about financial support for foreign students, while the photography shows students with blond hair, which is stereotypical for German students.
- The numbers and data about the scholarships are evidence that not all excellent students can be financed.

In other cases, the photography supports the utopian ideals (e.g. the blue sky as a symbol of the absence of pollution and of a good way of life). While photography apparently has characteristics of objectivity and reality, it is actually the result of a process of selecting and structuring the real world (see Brucculeri 2009: 17).

The interaction between verbal and graphic components seems to show the desire of the institution to find a balance between reality and utopia. Utopia is intended as a place to be reached gradually, and with concrete actions.

References

Agnello, M. (2013). *Semiotica dei colori*. Rome: Carocci.

Baglieri, S. and A. Shapiro (2012). *Disability Studies and the Inclusive Classroom*. New York: Routledge.

Barthes, R. (1973). "Texte (théorie du)". *Encyclopaedia universalis,* vol. 15. Paris: 996–1000.

Brambilla, M. (2012). "L'uso degli allocutivi in ambiente accademico in Italia e in Germania". In: Cinato, L., M. Costa, D. Ponti, and M. Ravetto. (eds) *Intrecci di lingua e cultura. Studi in onore di Sandra Bosco Coletsos*. Rome: Aracne: 65–78.

Brinker, K., H. Cölfen, and S. Pappert (2014). *Linguistische Textanalyse: Eine Einführung in Grundbegriffe und Methoden*, 8th Edition. Berlin: Erich Schmidt Verlag.

Brucculeri, M. C. (2009). *Semiotica del turismo*. Rome: Carocci.

Crestani, V. (2010). *Wortbildung und Wirtschaftssprachen. Vergleich deutscher und italienischer Texte*. Bern: Peter Lang.

Crestani, V. (2015a). "YouTube-Videoanleitungen im DaF-Unterricht: Vorschläge für Sprache-in-Interaktion-Kompetenzförderung". In: Imo, W. and S. Moraldo (eds) *Interaktionale Sprache und ihre Didaktisierung im DaF-Unterricht*. Tübingen: Stauffenburg: 279–300.

Crestani, V. (2015b). "*Busfahrer, trasporto merci, meat-eater*: Rektionskomposita im Deutschen, im Italienischen und im Englischen". *Germanistische Kontexte* 1, 29–50.

Crestani, V. (accepted). "Il discorso dell'alimentazione sostenibile in testi tedeschi e svizzeri: aspetti verbali e semiotici".

De Beaugrande, R. and W. U. Dressler (1981). *Introduction to text linguistics*. London: Routledge.

Gaeta, L. (2010). "Synthetic compounds. With special reference to German". In: Scalise, S. and I. Vogel (eds) *Cross-Disciplinary Issues in Compounding*. Amsterdam: Benjamins: 219–236.

Heinemann, W. (2000): Das Isotopiekonzept. In: Antos, G., K. Brinker, W. Heinemann and S. Sager (eds) *Text- und Gesprächslinguistik. Ein internationales Handbuch zeitgenössischer Forschung.* Berlin and New York: de Gruyter, 54–60.

Jewitt, C. (ed) (2011). *The Routledge Handbook of Multimodal Analysis*, reprint edition. London: Routledge.

Laskewicz, Z. (2004). "Towards Multimedia Textuality: new types of text in a changing world". <http://www.nachtschimmen.eu/zachar/writer/0404_TMT.htm>.

Miemietz, B. (1993). *Motivation zur Motion: Zur Bezeichnung von Frauen durch Feminina und Maskulina im Polnischen*. Frankfurt a. M.: Peter Lang.

Robustelli, C. (2000). "Lingua e identità di genere. Problemi attuali nell'italiano". *Studi Italiani di Linguistica Teorica e Applicata* 29(3): 507–527.

Schwarz, J. (1999). *Die Kategorie Weiblichkeit im Tschechischen: die Verwendung femininer und maskuliner Personenbezeichnungen für Frauen im Vergleich zum Deutschen*. Prague: Univerzita Karlova.

Siever, T., P. Schlobinski, and J. Runkehl (eds) (2005). *Websprache.net. Sprache und Kommunikation im Internet*. Berlin and New York: de Gruyter.

Spieß, C., S. Günther, and D. Hüpper (2012). "Perspektive der Genderlinguistik – eine Einführung in den Sammelband". In: Günther, S., D. Hüpper, and C. Spieß (eds) *Genderlinguistik. Sprachliche Konstruktionen von Geschlechtsidentität*. Berlin and Boston: de Gruyter, 1–28.

Sites

<http://studentenwerk.essen-duisburg.de> (Accessed October 2015)

<http://www.asu-arbeitsmedizin.com> (Accessed August 2015)

<http://www.spiegel.de/unispiegel/studium/studium-erstsemester-studenten-erzaehlen-ueber-heimweh-a-997128.html> (Accessed August 2015)

<http://www.umweltbundesamt.de> (Accessed August 2015)

<https://www.uni-due.de/> (Accessed October 2015)

<www.collinsdicionary.com> (Accessed August 2015)

<www.duden.de> (Accessed August 2015)

<www.nielsen.com> (Accessed September 2015)

Giampietro Gobo

The Care Factor: A Proposal for Improving Equality in Scientific Careers[1]

"Usually behind every great man there is a great woman;
but behind a great woman there is always a great domestic servant."
–Luciana Littizzetto, comedian

In academia, as in other spheres of work, there exist disparities associated with discrimination based on discipline. In academia, for instance, some research

1 I wish to thank Emanuela Abbatecola, Maria Carmela Agodi, Miriam Bait, Marina Marzia Brambilla, Diana Urania Galetta, Chiara Guglielmetti, Mita Marra, Antonella Nappi, Gloria Regonini, Paola Ronfani, Renata Semenza, Silvia Salini and Anna Lisa Tota for comments and suggestions.

topics are considered more mainstream than others, and some research methods more relevant than others. Discrimination also happens in terms of ethnicity and gender, etc. For some of these disparities, there is significant reflection literature written on the topic; but other forms of discrimination have come under less critical scrutiny.

An issue not yet sufficiently explored concerns the relationship between care activities and scientific research, particularly the question of how the former could have an impact on the latter. More specifically, how with increasing care-giving we can have a slowdown in scientific output. Moreover, since the latter is one of the main criteria for hiring or promoting a candidate, it becomes especially important to shed light on this relationship.

This correlation does not, at first, pertain to the theme of the genre, but who (man or woman) is engaged in care (both motherhood and fatherhood). In the younger generations, care activities are further balancing out between the genders, albeit slowly, with men assuming greater responsibilities and demands in the domestic sphere.

However, since they still expect women to be more engaged in the care, this correlation is particularly unfavorable to women. For this reason, practical proposals to reduce inequalities in scientific careers are urgently needed.

1. How is the Scientific Product of an Individual (Currently) Evaluated?

Because publications are an increasingly important criterion for evaluating a candidate who is applying for a job or being considered for a promotion, it is important to carefully analyze the inner mechanisms of the evaluation process. This provides insight into how these evaluative mechanisms tend to discriminate, often unconsciously, against the very people who are most involved in care activities, which (we assume) is the main cause of the slowdown in the production of scientific research. In the following sections, we will analyze this evaluation process.

Measurement?

Usually the "measurement" of the scientific production refers to the output of an individual. In fact, measurement is an evaluative operation allowed only

a. with variables that have *continuous* properties (e.g. time, weight, height, income, age etc.); and
b. where there is a *measurement* unit (based on a predetermined amount or size, which is conventionally accepted).

Only these two requirements, which must both be present, allow for measurement to be used (see Marradi 1981, 1990 for a discussion).

Instead, the evaluation of scientific production rests on other procedures or methods, including counting, classification and reading (the content of a publication). None of these methods, however, are measurements (see Tab. 1).

Tab. 1: Three procedures used in the evaluation of scientific products

Counting	**Classifying** (not countable)	**Reading** (the content of the scientific product)
• **h-index** (based on the number of quotations) • **number of publications** (above or below the median) • ***Impact Factor*** of the journals in which the article was published • **number** of readers and downloads for a publication • **number** of tags, bookmarks, comments, tweets or blog posts to assess the impact of authors or publications	• the **language** (preferably English) in which s/he published • if it is **books** or **articles,** or **chapters** (of books) • whether published with **editors** (in the case of publications on journals) • **taxonomy** of the journals (top, average, bottom journals, with preference for the top ones…) • **taxonomy** of publishers (with preference for prestigious ones…)	• thorough reading • fast reading • skimming

The contemporary tragedy is that counting and classifying are rapidly replacing reading, which should be the main evaluative method used. In other words, how do I evaluate an article if I have not previously read it? How can I assess the scientific output of an individual if I have not read their work? Bibliometric and classificatory *nouvelle vague* has invented shortcuts: just a few counts and classifications, and you are done. In this way, two (useful but surely) peripheral modes that do not focus on the content of a publication have now become the main evaluative procedures. This results in a loss of the possibility to evaluate the **merit** of a publication[2].

2 This is precisely the criticism contained in the Declaration on Research Assessment (DORA), drawn up by a group of editors and publishers of scientific journals, gathered in San Francisco, California, on December 16, 2012 at the annual meeting of the American Society for Cell Biology (ASCB). Since then, this declaration has been signed

The situation is such that evaluators and recruiters are often flooded with an avalanche of publications that must be *evaluated in a short time* (an oxymoron!) in the midst of all the many other things a scientist has to do. The result is that candidates' scientific publications are not read closely, except for a few cases, with evaluators only skimming to search for the soundness of scholarship. This is, in short, a humiliating practice for the authors of these publications.

2. Monitoring or Evaluating?

This reflection on the (current) assessment practices of the scientific research published by a candidate is that perhaps what evaluators and recruiters are doing is not a genuine evaluation, even if the intent is honest. In fact, evaluation should be "a cognitive activity [that] aims to provide a judgement of an action (or set of coordinated actions) performed intentionally/or being undertaken, designed to produce external effects, using the tools of the social sciences, according to strict and codified procedures" (Palumbo 2001: 59). If we accept this definition in the context of the evaluators' scientific community, counting and classifying (the first two procedures) should not be fully included in the evaluation because the result of these methods is not (subjective) judgment, but simple data that could be provided by any person (including an administrator!) who has been given instructions and a spreadsheet for ranking criteria like journal quality or publisher ranking. In the third method of reading, however, the expert, whether a scientist in the discipline or research area is indispensable to judge, discern, identify, and understand the content of the publication.

Unlike the first two procedures, there is little evaluation and much **monitoring**. The latter, like basic research, applied research (Palumbo 2001: 64), auditing (Bezzi 2001: 65, 67), benchmarking, certifying, and social budgeting, is **not** evaluation. Indeed Bezzi (2001: 66) states that monitoring is the opposite of evaluating, which includes the tasks of monitoring and auditing, but is not limited to them.

by hundreds of organizations (scientific journals and associations) and thousands of scholars. Joining DORA commits to supporting the adoption of scientifically correct research practices evaluation. DORA contains 18 recommendations addressed to the different actors of the research world: funding agencies, institutions, publishers, organizations producing bibliometric data and individual researchers. One of them insists on eliminating the use of metrics related to journals – like Impact Factor – for funding, recruitment and promotions (or) as a surrogate measure of the quality, or to evaluate the contribution of the individual scientist, or decisions relating to recruitment, promotion and funding.

Evaluation goes further because it adds critical judgment. Counting and classifying can therefore only be a pre-condition of the evaluation, an initial screening and not an assessment itself.

3. Against Abstractive Evaluation: Desperately Seeking Society

The two procedures (that have become dominant in the evaluation of scientific products) fall into an abstracted evaluation pattern, divorced from the social dynamics, epistemologically naive, politically inexperienced and not reflective. This pattern, ostensibly rational and transparent, does not take into account how indicators are (socially) constructed, and which representations, mental models and tacit knowledge are embedded in them.

What Makes a Person a Good Researcher: A Problem of Conceptualization

All methodological textbooks teach that in order to conduct research, it is first necessary to conceptualize the phenomenon, i.e. defining the so-called 'object of research'. One cannot start researching family or poverty, for instance, without first defining what a family is (and we know how definitions are controversial and change over time) or what poverty is. In other words, what criteria I adopt include that particular relationship between people in the category of 'family', or what requirements a person must have to be defined as poor.

The same applies when it comes to evaluating a researcher. **Before** evaluating a researcher, you must conceptualize who (or what) is a good researcher. **Before** choosing performance indicators, we should discuss what the attributes (according to constructivism) or properties/characteristics (following realism) of the concept of 'good researcher' (the so-called 'intention') are. Instead, the common practice is actually moving in reverse: indicators are chosen in a confused, abstract and naïve way without reflecting on what is behind these cognitive tools or what their cultural background is. This information is then used to form the concept of what a good researcher is.

An Evaluation Without…Society

Existing tools for the assessment of scientific production measure people as if they were impersonal databases, not social entities. These tools forget that they are people with biographies and social trajectories etc. It is as if the authors and their products belong to two separate and unconnected worlds. A (welcome) opposite tendency is represented by the think tank New Economics Foundation (Nef), a group of 50 economists famous for bringing into the agenda of the G7 and G8 issues such as international debt, whose motto is (significantly): "Economics as if people and the planet mattered."

Starting from the assumption (therefore theory) that there should be a direct correspondence between what we are paid and the value our work generates for society, Eilis Lawlor, Helen Kersley and Susan Steed (the authors of the Nef report) calculated the economic value of six different jobs, three paid very well and three very poorly. As the authors explain in the introduction:

> We take a new approach to looking at the value of work. We go beyond how much different professions are paid to look at what they contribute to society. We use some of the principles and valuation techniques of Social Return on Investment analysis to quantify the social, environmental and economic value that these roles produce – or in some cases undermine. (4)

Following the criterion of linking salaries to the well-being that a job brings to the community, Nef concludes that bankers abandon society and cause damage to the global economy. Under this logic, when comparing a garbage collector to a tax affairs lawyer, the former contributes with his work to the health of the environment through the recycling of garbage, while the second harms society because he comes up with schemes to enable his clients to pay fewer taxes. So, when looking at the social contribution of their job, it turns out that the kinds of jobs that are paid less are actually the most useful to society.

This example shows how the assessment is primarily a social and political practice, guided by certain "theory-driven", cultural assumptions (whether tacit or explicit) and by particular mental models of what makes (in the case of scientific careers assessment) a good researcher. It is not a simple observation based on neutral formats and criteria. As Chen and Rossi (1981, 1989), Chen (1990), Weiss (1995, 1997), Pawson and Tilley (1997) and many others have long pointed out, assessment is only secondarily a technical issue.

4. The Tacit Assumptions of the Academic Evaluation: the Hidden Cultural Model

Given that evaluation is primarily a theoretical activity, it is worth trying to explore the main cultural assumptions (tacit and/or explicit) and the particular mental model of what makes a good researcher. In academic common sense, an ideal researcher is one who:

- Teaches a lot and delivers high-quality teaching.
- Publishes extensively, particularly scholarly or prestigious publications, but at least peer-reviewed articles and books in innovative areas of research.
- Conducts good research.
- Wins national and international grants.
- Participates in national and international conferences.
- Accepts institutional duties.
- Participates in the intellectual life of the department in the form of seminars, conferences, etc.

Any academic would subscribe to the responsibilities outlined on this list (Keith and Moore 1995; Pescosolido and Aminzade 1999; Golde and Walker 2006; Sweitzer 2009).

Society Enters Evaluation: Assessment as If People Mattered

Ask ourselves now: who could perform better (equal intelligence) in all of these areas? A sociologically plausible ranking would be:

1. Single
2. Person with partner, without children
3. Person with partner, with 1 child
4. Person with partner, with 2 children
5. Person with partner, with 3 children
6. etc.

Intervening social factors (like having to care for a sick or disabled person, having a partner away for work, having the support of grandparents, or having access to financial resources to pay for a babysitter etc.) must be left out of consideration in this case because otherwise it becomes difficult to manage the complexity and evaluate it with standardized instruments.

Obviously, we are talking about people of both genders here, not men or women in particular. For the moment, gender does not come into play. And, when it does, it will make the evaluation even more social.If this ranking is sociologically plausible (and we will soon see how it is) other questions arise:

- What is the cultural model underlying these evaluation criteria?
- What tacit assumptions are embedded in it?
- What is the underlying profile?
- Maybe that associated with Nobel Prize winner Rita Levi Montalcini (single) or the famous astrophysicist Margherita Hack (partner, no children)?

- Is it reasonable to assume that researchers who have had children have also experienced a drop in scientific production, defined as reduced capacity to conduct research and guarantee an institutional presence?

Female Nobel Prize Winners

An interesting example to test this hypothesis is using the Nobel Prize award, particularly because only a few women have been awarded the Nobel Prize (see Cole 1987, Wade 2002). In fact, the Nobel Prize has only been awarded to women 47 times between 1901 and 2014 (only one woman, Marie Curie, has been honored twice). During this same period, 814 men were awarded Nobel Prizes. Women's creativity is clearly underrepresented in science.

But even more interesting for our hypothesis is the fact that many of these women did not have children (Stemwedel 2009). By comparing the 11 female Nobel laureates in physics, chemistry and physiology/medicine between 1901 and 2006 with 37 males who received the Nobel Prize in the same area one year prior and one year after them, it was found "that female Nobel laureates were significantly less likely to marry and have children. When female laureates had children, they had significantly fewer children than male laureates. Female laureates also had fewer publications than their male counterparts" (Charyton, Elliott, Rahman, Woodard and Dedios 2011: 203). These authors conclude that eminent women scientists tend to choose the pursuit of scientific discovery over starting families more frequently than prestigious male scientists.

If this were to hold true for the vast majority of women working in the academy, would it be reasonable for women to ask for a correction, a weight, an adjustment to the current assessment procedures that takes into account the number of children and the demands of care giving?

5. The International Context and Southern Europe

Growing literature on the topic and other research shows that at least in Southern and Eastern Europe, fathers and mothers are not yet equal in their performance.

Weekly Hours

Among Europeans, Greeks work the most of all countries, with an average of 42.2 hours per week (British Office for National Statistics, 2012). The Germans, on the other hand, work only 35.6 hours. These two countries are at opposite ends of the current crisis in Europe. The hours worked and economic performance seem to be upside-down, with the hardest workers in Greece, and almost 'lazy' workers in

Germany. In Bulgaria, Czech Republic, Poland, Romania, and Slovakia, people also work on average more than 40 hours per week. Denmark is even more extreme than Germany, at not even 34 hours, and Scandinavian countries in general work less than the classic 40-hour week.

One explanation for this could be the difference in productivity: Germany is dominated by well-organized companies and modern technologies, financial opportunities, supply chains and advanced distribution, and low corruption. At the other extreme is Greece, where the people work longer with less achieved. In 2012, in a measure of productivity in the European Union, Germany was at 123.7, while Greece landed at 76.3.

It must be acknowledged that it is not the number of hours worked that creates wealth and efficiency. We must understand why in many countries, workers but especially managers stay so long at the office, while women tend to concentrate their presence at work in order to take care of other domestic and family responsibilities. The reason behind this typically male behavior is that many companies' people construct their career through relationships instead of on merit alone. Staying late in the office creates solidarity, especially among male workers because most of the women have already gone home. And in organizations where you advance through cliques, membership and loyalty are decisive factors. This is not true everywhere. But certainly at companies where workers put in long workdays are not necessarily focused on quality and productivity. Taino (2012) reports some statements of Anglophone managers: "When the clock strikes five, the team goes home," says Max Cameron, CEO of Big Bang Technology, a hi-tech Canadian company "I tell everyone that the game is over and they lost". Cameron believes that the mentality of the hero who sacrifices himself for the company is highly damaging to the company. On the one hand it leads to burnout, a mental disorder of those who work under stress for too long. On the other it creates an environment where no one can become a leader without sacrificing his life, who then cannot be promoted even if s/he has the qualities to be so. As Cameron says, this "erodes the leadership".

Barry Sherman, the CEO of the American company Pep Productivity Solutions, says that the non-economic reasons why someone works overtime are twofold: "The first is that s/he does not want to go home and the second is that s/he does not know how to work effectively" (Taino 2012). People who work overtime probably do it to please their boss and advance in their career, but also because the external environment permits it. In Sweden, for example, staying at the office beyond the typical end of the workday is considered objectionable from the point of view of society; it is seen as a selfish choice and suggests that that person has no interest in others or the outside world, beginning with the family. This cultural belief is so dominant that only 1 percent of Swedes work more than 50 hours per week (OECD 2012) and overall men on average devote 177 minutes per day to cooking, washing and caring for children, compared to 249 minutes for women. In Italy, however, men spend 103 minutes for activities related to the family compared with 326 minutes for women. Career and social relations both inside and outside the office, therefore, explain many of the reasons for overtime.

In 2002 in Sweden, the Toyota company in Gothenburg, the second largest city in the country, switched to a six-hour working day, resulting in happier employees (in a country where winter depression affects people more than elsewhere), a lower rate of turnover and an increase in profits. This experience is also expanding in other companies in Sweden, where many employees have set themselves the goal of accomplishing more work in a shorter amount of time so they can have more time to devote to their personal lives. Filimundus, an application developer company based in Stockholm, also introduced six-hour days in 2001. "The eight working hours are not as effective as you think," says Linus Felds, CEO of the company. "To remain fixed on the same work for eight hours is difficult. To do that, we usually intersperse work with breaks".

Japan is a country where merit is poorly rewarded and networks of relations prevail, and where there are still high barriers between male and female workers. According to the trade unions, on average, one supervisor out of ten leaves the office after 11 pm, and 84 percent of workers generally work beyond the established working hours. To this point, Japanese has the word *karoshi*, which means death from being overworked. Similar situations in which workers stay at the office until the boss leaves are seen in South Korea and Singapore. Then there are those who work too much, leading to organizational dysfunction. In this case, the problem is serious for the company; it may be getting lots of overtime from its employees, but the trade-off is quality and creativity. There are also serious problems for workers: a 2012 study at the Institute for Health and Employment of Finland discovered that those who regularly work more than the classic eight hours per day had a 40–80 percent increased risk of having heart problems and ran a greater risk of developing dementia due to "prolonged exposure to stress" says Marianna Virtanen, who led the research.

Productivity and Prolongation of Working Hours: Two Variables… Inversely Proportional

A comparative analysis of dozens of companies commissioned in 2014 by the Californian company Seagate Technology reveals that managers and employees waste a good deal of time each day sending and replying to emails and participating in meetings. They "really" only work a few hours, and yet the time is never enough.

In one American multinational, for example, employees spend on average 20 hours per week in meetings and send or receive 3700 emails per year. A typical manager of a company this size "consumes" 400 hours a week of its employees' time in email correspondence and meetings. "The meetings are frequently the realm of multi-tasking," quips an expert, "people spend their time writing messages on their mobile phones, in practice not even listen[ing] to what others say" (Franceschini 2014). In addition, meetings almost always last too long: when a meeting exceeds 90 minutes, attention wanes and participants lose sense of the meeting objectives. These bad practices are especially detrimental to the careers of people who have very tight timelines at work due to their child-care commitments.

The Double Burden

In the 1970s, the first reflections on the concept of "double presence" (Balbo 1978) or "double burden" "double day", "double duty", "second shift" (Hochschild and Machung 1990) appeared. These expressions indicate the dual roles shouldered by women: public and private, reproductive in the family and productive in society. Under this concept, a woman is squeezed between her dual responsibilities both to the family and to her independence, represented by the work, which has a negative impact on her. This phenomenon is found in all companies and continents, naturally in different contexts and intensities. In fact, men's contribution, in order to alleviate double burden, is understood as an option for most couples, even those who share family responsibilities equally…the male role in the allocation of family work is limited to a minimum necessary assistance (Jana 2011: 176–177).

6. Social Policies to Reduce Gender Inequalities

Worldwide since the late 1970s, there have been numerous public policies launched in favor of gender equality. There are many different policies, some of which are potentially complementary, others ideologically incompatible due to the distinct cultural and theoretical perspectives that guide them. Let us describe these policies briefly (for more see Marra 2014). The first group of policies is focused on promoting affirmative action. First borne in the United States, they aim to overcome specific inequalities between men and women, such as instituting quotas for the boards of listed companies or universities, which aims to ensure the autonomy of women not only in terms of social distress but also (and especially) in the higher levels of the socio-economic and political-institutional organization of an economically advanced democracy (Shalev 2008; Marra 2014).

In contrast, in continental Europe policies in favor of women have resorted to social protection (i.e. services provided by the institutions of the welfare state, monetary transfers as maternity or illness allowances, retirement etc.) rather than the opportunities for emancipation and economic growth that the labor market could offer.

According to this second approach, it is necessary to remove "the unequal division of extra-work duties and care-giving, the inadequateness of reconciliation services (whether public or private) and the resistance by enterprises and public administrations to rethink the models of work organization" (Marra 2014).

A third group of public policies appeared in the middle of the 1990s, brought about by the European Union and international organizations. It refers to the approach known as gender mainstreaming, which originated in Northern Europe (Rubery 2002; Verloo 2005; Gornick and Meyers 2008; Knijn and Smit 2009). It argues that, since there is not a biological relationship between gender and social roles and responsibilities, the existing differences between the genders can be filled in order to make the tasks in the two main social spheres of work and family interchangeable and assimilable. This mutual replaceability would allow for the development of a more egalitarian society. According to this approach, it is necessary to remove "the unequal division of extra-work duty and care, the inadequateness of the reconciliation services (whether public or private) and the resistance of enterprises and governments to rethink the models of work organization" (Marra 2014).

Finally, a fourth group of proposals refers to more purely feminist approaches (Pillow 2002; Sielbeck-Bowen et al. 2002) that are sometimes in conflict with the gender approach, which is accused of promoting an unconscious acceptance of the capitalist model of production, tacitly embodied in the public policies they propose. Instead, feminist scholars recognize the importance of caregiving, social reproduction, and appreciate the differences women make in lifestyle choices. For them, the differences between and within genders are irreducible and an asset to society, a resource to respect and promote balancing the pressures to homologation embodied in the social organization of the most advanced economies (Marra 2014).

These four approaches are often difficult to reconcile, so much so, in fact, that it has been suggested to move beyond the feminist and gender-based approaches (Marra 2014). Let us take an illustrative example: in order to combine work with caring for young children, a society may enact different public policies including increasing available places in kindergartens and nursery schools; providing monetary support and contributions to help families who decide to send their children to these institutions; reducing school fees; and providing supplementary payments to cover the costs of babysitting or creating other services.

However, this logic is very functional within a model of capitalist production. What if a parent prefers not to make use of daycares or babysitters during a child's first three years of life? In any case, what is better for a child than to spend the first years of life in close contact with (at least) one parent? Why impose institutionalization from the very first years of life for an infant? Why entrust their education to babysitters who often come from countries with highly sexist cultural models? Why abdicate our own educational tasks to grandparents or other relatives, who perhaps do not share the same educational models as those the parents wish their children to have? Due to these reasons, the balance of the burden of responsibility for family care through new welfare benefits and flexible forms of work organization cannot be tailored to suit everyone. A much better approach is the increasing number of colleges and universities that have instituted (over the last decade) a variety of policies for new parents including tenure clock extensions, reductions in teaching duties, and parental leave, to name just a few (Ward and Wolf-Wendel 2012).

However, greater equality between men and women would probably be achieved if the tasks of care were equally distributed between genders. Unfortunately, this is (still) not not a reality for most couples. Therefore, mother-scientists can hardly compete, in terms of scientific production (journalistic), with their male colleagues, whether or not they are themselves fathers. And, often, these women cannot make it.

Can We Think of Something Faster?

Several studies (Barclay and Lupton 1999; Harrington, Van Deusen and Humberd 2011; Miller 2011; Jana 2011; Hook and Wolfe 2013; Kaufman and Bernhardt 2014; Rehel 2014, Pizzorno, Benozzo, Fina, Sabato and Scopesi 2014) show that the traditional division of gender roles in child rearing is changing. A new model is emerging in which there is greater equality in managing a couple's double careers: more and more fathers participate in family life and are involved in care-giving activities (Marotte, Reynolds and Savarese 2011). In addition, social and conciliation policies (Marra 2012), which help to rebalance care-giving activities between gender roles, are increasing. This is seen, for instance, in the increase of flexible regulation of work hours both for men and women; the increase and improvement of health services; and better reconciliation practices for women and men in businesses, trade unions and at the state level in civil society (see Gasauka 2011). However, these changes are slow and the effects of these policies take decades to manifest themselves.

Instead of waiting for these effects to take root, could we immediately introduce some correctives to reduce (at least partially) the existing inequality, particularly in the scientific production of the mothers, which (as we have seen) is often the first criterion used to evaluate a scientist or academic professional?

7. For a Care-Sensitive (and Mother-Sensitive) Assessment

If care-giving results in a slowdown of scientific production, then we need an evaluation practice that takes this into account..However, because men and women currently do not participate equally in the child-rearing and family responsibilities, it becomes necessary that the assessment should be geared more towards the mothers. If men and women are different (and often unequal) in society, we cannot assume that the effects of this diversity (and inequality) are simply suspended when we turn to scientific production. The same applies to mothers. There is however currently little attention given towards providing a differentiated assessment of scientific production, which essentially assumes that men and women, fathers and mothers, are or should be equal.

Unfortunately, for Mothers … Little Data

Researchers who compare the scientific output between men and women to see if there is a real difference rarely report whether these women under study have children. There is little academic data that includes this information and has the related variable that would be useful for a more accurate assessment.

In fact, when it comes to scientific output, it does not seem reasonable to hypothesize differences between men and women if both are childless. Instead, the discussion would become more interesting if we could ascertain whether children (in addition to illness, providing care for parents, lack of livelihood etc. which here we do not consider) are a possible and important cause of the slowdown in scientific output and contribute to difficulties in doing research and participating in the institutional life of the department, conferences etc. which, again, we do not consider here. Currently there are many studies conducted in different countries which show that child rearing is still strongly attributed to the mothers.

Unfortunately, current research and comparisons rarely outsource this data and therefore do not help us to understand whether children have an impact on women's scientific production. For example, Tower, Plummer and Ridgewell

(2007) conducted a study on six of the top journals in the world, as rated by the 2006 Thompson ISI index. They chose two for each category: two in science (*Science* and *Cancer Journal for Clinicians*), two in business (*Academy of Management Review* and *Quarterly Journal of Economics*) and two in the social sciences (*Archive of General Psychiatry* and *Harvard Law Review*). Leaving aside the questionability of this selection (namely, I am not sure how many social scientists know of the existence of the last two journals), they found no difference in productivity when the percentage of the women participating in the academic work force is factored in: women comprised 30–35 percent of participation rates in academic university positions and represented almost 30 percent of the authors in the top-tiered journals. In addition, they did not find any significantly statistical difference in journal Impact Factor ratings between women and men. This is an example of abstractive statistical analyses, which are totally decontextualized. Because their analysis only covers six top journals, the generalizability of these results should be taken cautiously, and the authors do not check whether those women have children. They only take into consideration the (abstract) variable of gender without providing any contextual analysis to understand the biography of these women and men (e.g. age, marital status, children, etc.). The same problems are to be found in the studies by Rothausen-Vange, Marler and Wright (2005) and Dasaratha, Raghunandam, Logan and Barkman (1997).

So there only remains the option of looking at comparative research between men and women, and then weighing these results using a virtual or "thought experiment" (*Gedankenexperiment*) with the use of "counterfactual" conditionals (Van Dijk 1977: 79–81), a type of research used in economics, physics, cognitive sciences, history, etc. (see Gobo 2008: 151–152).

Women's Scholarly Productivity

Much of the literature on work and family issues in academia suggests that women with children have a harder time maintaining an ideal career because of the difficulty of combining work and family activities, both of which are regarded as "greedy institutions" (Hochschild 1975). Women "are expected to (and often do) take on more child-rearing and housework responsibilities. If separated or divorced, women are more likely to be the custodial parent. There is considerable literature that women academics are hampered in their efforts to have an ideal career" (Spalter-Roth and Van Vooren 2012). According to a report from the Committee on the Status of Women in Sociology (2004): "Women may face serious disadvantages. Careers often are built … around a model of a worker who has no competing responsibilities to work and is able to devote full attention to (usually his) professional life. Persons who do not conform to this pattern of the unencumbered worker will be disadvantaged in achieving success within the profession".

In a study of doctoral students at the University of California, over 70 percent reported that they considered academic careers in universities unfriendly to family life (Mason 2012). Women with children "may be unable to regularly stay late to muse over intellectual questions with colleagues at the office or a local pub, but instead may have to pick up children from school or daycare or return home to prepare dinner" (Spalter-Roth and Van Vooren 2012). In addition, women may sometimes need to bring the baby to class with them (Kennelly and Spalter-Roth 2006).

Research suggests that parenting within the academy is a gendered phenomenon. Mason and Goulden's (2002) widely-cited study of a nationally representative sample of PhD recipients between 1973–1999 finds that raising children, especially early in one's academic career, has a negative effect on women's careers. However, this was not true for men. Women who have children are more likely than men with children to have marginal or alternative careers.

However, the research on women's scientific productivity offers other controversial results, which are not always easy to interpret. Kyvik (1990) argues, on one hand, that women become more productive when the children get older, because the children are more independent and less in need of care. In the same article, amazingly enough, Kyvik also states that both men and women, married and divorced people are more productive than singles.

The latter statements are not credible in the light of the the former statements. For example, the following: women with children are more productive than women without children (Kyvik 1990). These statistics are out of context, without accounting for social dynamics. In other words, it would be important to know: who are those women with children? How many children do they have? Do they have domestic help? To which social class do they belong? Without this information, interpretation appears shaky.

According to the study by Long, Allison and McGinnis (1993), although men and women start out as assistant professors with similar levels of productivity, after six years men have significantly more publications. Kyvik and Teigen (1996) observe that in the span 1989–1991 (of their database), men had published an average of 6.9 articles, while women had only published 5.6 (20 percent less). During the same period (1989–1991), male faculty members under age 40 published twice as many article equivalents than their female counterparts, whereas for faculty over age 40 the difference is small (10–15 percent) (Kyvik and Teigen 1996).

From this research, some clarity begins to emerge on the differences between men and women.

Ward and Wolf-Wendel (2012) conducted a longitudinal study, interviewing over 100 women who are both professors and mothers, examining how they navigated their professional lives at different career stages. They studied how tenure-track women faculty members managed work and family in their early careers (pre-tenure), when their children were young (under the age of five), and then again in mid-career (post-tenure) when their children were older. The findings suggest that family plays a role in how people develop in their academic careers, just as careers play a role in how people evolve within their family.

Women and Bibliometric: What Happened in Italy?

In 2012, in Italy, the National Agency for the Evaluation of the University System and Research (ANVUR) settled the minimum requirements to become a full professor with a ministerial decree. Shortly after, Corsi and Zacchia (2013) did a simulation by applying the ANVUR's bibliometric 'recipe' to the scientific output of female economists (full faculty members as well as potential candidates for promotion to full professor), to see how many of them satisfied the criteria established by the ANVUR. The results were surprising.

If we look at the median number of journal articles and book chapters, out of a total of 301 female economists (including 110 associate professors) only 22 percent of lecturers and associate professors satisfied the first requirement, which had a median equal to or greater than 8. Unlike for men, the percentage of success was 35 percent. If we look at the median number of books published, only 3.6 percent of female lecturers and associate professors had published at least one monograph over the past decade. In this case, the percentage achieved by males was higher than or equal to 9 percent.

Finally, if we monitor the median number of publications in top journals, the criterion of excellence of the economic disciplines SECS-P01/P06 ranges from 0 (in science of finance, economic history, history of economic thought) to 6 publications (Econometrics) in ten years. Although this was poor coverage of the top journals in the Econlit database, only 26 percent of female economists had at least one publication in the past decade included in the list of the requirements of "excellence": specifically, 25 percent of associate professors and 27 percent of lecturers. In this case, the gender gap was more pronounced because about 90 percent of men had at least one article in the last ten years published a top journal.

Causes of Gender Disparities in Academic Publishing

In literature, the underproduction of academic women in research outcomes has been traced to the following root causes:

- Women and men tend to collaborate with co-authors of the same sex; because there are relatively few women in faculties, women have more difficulties finding co-authors (Ashcroft, Bigger and Coates 1996; Suitor, Mecom and Feld 2001; Bentley 2003).
- Females are more likely to work in non-tenure track, part-time and temporary positions, or work at teaching colleges, leaving less time for research and publishing (Dasaratha Raghunandam, Logan and Barkman 1997; Mathews and Andersen 2001; Robinson 2006), more involved in service activities at the expense of research (Dasaratha, Raghunandam, Logan and Barkman 1997; Maske, Durden and Gaynor 2003; Corley and Gaughan 2005; Robinson 2006) and disadvantaged by family responsibilities. Men spend more time at the university and less time at home, even among married faculty, especially during child-rearing years (Mathews and Andersen 2001; Bentley 2003; Suitor, Mecom and Feld 2001).

These factors are slowly changing. However, change is happening gradually, and social and cultural changes are not easily predictable. Therefore, it is necessary to do something now.

8. Contextualizing Indicators (and Consequently Factors and Indices)

To accelerate the achievement of equality in scientific careers, it is necessary to adopt different criteria for evaluating CVs, particularly for scientific output. If, as the literature has documented, men and women indeed face different realities (and diversity management is now a reality), it is not clear why they should be treated as equal.

To this end, the proposals can be various and diverse. If a candidate is strongly committed to child-rearing, their scientific production could be evaluated in the context of how many children they have[3]. That the weighting can be reasonable

3 Obviously taking into consideration that the number of children owns cardinal properties only apparently: in fact three children are not 3 times 1 son. Again social dynamics should be taken into account, because the burden of caring for three children depends on many factors: whether there are twins, how many years apart the children

is testified by the practice (now widely accepted and published) of normalizing the scientific production for the age of the candidate. So we can assume different remedies (even standardized as weights, corrective coefficients, adjustments, normalization etc.) that take into account the social dynamics and inequalities on behalf of those engaged in activities of care, in order to better assess their scientific production. For example, this could take the form of the following methods:

1. To normalize, to attribute a score, an additive weighting etc. to those who have children.
2. To give priority to the quality of publications, rather than quantity. Candidates could indicate three publications they consider their best, their most innovative, etc. and the assessment will be conducted on those only. So at least the referees would read them, which they do not do when they receive 20 publications.
3. To give space on the CV to quality management (practical skills, multi-tasking, negotiation skills, ability to reconcile different commitments etc.) that come from playing a role of a mother/father strongly present in the family (see 11).
4. To make a multidimensional assessment of research (see Tucci, Fontani and Ferrini 2010).

The Multidimensional Assessment of Research: The R Factor

In proposing a multidimensional evaluation of research, Economists Tucci, Fontani and Ferrini start from two very "social" assumptions (2010: 107):

1. The publication of articles is only one aspect, albeit an important one, of a researcher's scientific activity;
2. There are a number of activities, not always visible (and not always transformed into articles and citations), which nonetheless contribute to scientific progress.

To take better account of these two assumptions, they construct what they call the "index R-factor", which consists, in turn, of the following sub-indices:

- Articles published in journals
- Monographs and essays
- Grey literature
- Coordination activities (conferences, research groups, coordination, doctoral classes, theses supervision)
- Dissemination activities (seminars, conferences etc.)

---

are from each other, etc. This is why when society also enters into the equation, mathematics will be (welcome, but) always too late.

- Type activities publishing (journal editor, board member etc.)
- Administrative activities (dean, chair, coordinator, director of research centers).

Although not mentioned by the authors, additional information could also be included:

- Number of teaching hours
- Annual number of exams
- Number of theses and dissertations supervised.

As we can see, the term "evaluation" (attributed to this proposal) is very stretched, being nothing less (and no more) than a complex monitoring. However, the proposal looks very interesting and fruitful. In addition to these indices, a "care-factor index" would also be included, built on:

- Number of children
- Age of children
- Health status of children.

While taking into consideration other indices (such as parental capital, economic capital, etc.), even if significant, it could be complicated.

9. The C Weighting and Its "Enemies": Men and Women

Although there is broad consensus on the need to balance parental roles, when we move to operationalize this need through technical proposals, various opposing claims arise.The main "enemies" of the C weighting are primarily men (especially in the Latin countries, where men are generally more reluctant to split care practices with their partner) who reductively see this policy as an exclusive advantage for women. In fact, it is difficult to make it clear to men that this policy could also be applied to a father who decides to spend more time with children and family. It is no coincidence that, although in many countries there are rules allowing fathers to take advantage of parental leave (for child-rearing), requests for such leave are relatively rare: many men are ashamed to express this desire and prefer to give up the benefit rather than be exposed to jokes and macho criticism by males (and perhaps also by some female colleagues).

A second aspect concerns competition between universities: if the goal of a university is to maximize its results, it will tend to recruit candidates who publish more, do more research, have more education, are more institutionally present etc. Why should it hire or promote career advancement for those who are probably less productive? However, this type of reasoning (however widespread it may be) is based on limited rationality, unable to think globally, because ultimately the universities stand on (both economically and educationally) students. Moreover, the children of today could be the students of tomorrow. If for someone the reproduction of the species cannot be a positive value for the community (therefore as such not be positively evaluated), the fact remains that someone has to take responsibility for this task. Of course, we can decide to discourage reproductive activity among scientists and delegate it to the unemployed, the poor, migrants etc. However this (aberrant) plan should be explicit, and not tacitly activated in recruiting practices.

It may seem paradoxical, but there are already several generations of women in front of the C factor (as they still are or have been for years due to the affirmative action). One such example is the "wonder women", who were famous 1940s comic characters. These mothers who spent their lives doing somersaults balancing work and family and in the face of enormous personal sacrifices "made it", and became professors. They did not believe that having children would make them less productive. Indeed, they thought that if they made it, then other women would too, entering into a macho loop that damages younger generations of women (many of whom have raised the age of motherhood or chose not to procreate). Therefore, they believe that mothers do not need affirmative actions to win the men's competition.

10. Conclusions

Several research studies on care-giving highlight how it has had an impact on a scientist or academic's productivity, causing a related slowdown. The latter could prove to be detrimental when the candidate takes part in recruiting or promotion processes, as the number of publications is often used as an important criterion in evaluation.

Reconciliation policies are certainly a useful tool to help dilute this effect. It is also important to promote family-friendly cultures, environments and workplaces. Institutions of higher education are increasingly recognizing that being family friendly is an asset in terms of recruiting and retaining top faculty members (Evans and Grant 2008; Ward and Wolf-Wendel 2012; Mason, Wolfinger and Goulden 2013). However, they require social and cultural changes that are (unfortunately) not immediate. So practical proposals in the short and medium term to reduce inequalities in scientific careers are urgently needed. Affirmative action examples include company laws in which one third of the members of boards of directors of listed companies and publicly owned corporations must be women. There are also short and medium-term affirmative action policies (e.g. valid for 10 years) within which it is hoped to achieve the goal of removing the obstacles that have so far limited women's access to leadership roles, encouraging a process of cultural renewal in support for greater meritocracy and growth opportunities. Through these actions thousands of women had (and continue to have) the opportunity to take on leadership roles. That said, it would not hurt to extend this rule to academic staff such as the university senate and the boards of directors of the university.

One of these affirmative action policies in the academy could be the use of the Care Factor, a tool to weigh the scientific productivity of a candidate who is involved in child-rearing. It is a transitional instrument, certainly not permanent, but useful to balance the gap between those who are involved in care activities and those who are not.

However, the Care Factor should not be conceived of as a proposal that rewards those who care for children. It is not meant to reward those with care-giving responsibilities, but these people should also not be penalized. The care activity should be enhanced even further and become one of the different criteria for recruitment and promotion. In fact, child-rearing is not to be conceived exclusively as a burden, a responsibility outside the realm of academic activities. Unfortunately, as Ward and Wolf-Wendel (2012) show, much of the existing literature on balancing work and family presents a pessimistic view and offers cautionary tales of what to avoid and how to avoid it. In contrast, child-rearing brings sorely needed skills into the academy, like every other working sphere. As Balbo (1978) theorizes, the "double presence" is a way to "pass through many worlds" and thus be more innovative in both work and family. Competences and practical skills learned from juggling tasks in many areas, from negotiating and reconciling different needs to mediating between different instances, come in handy especially in collective dimensions of research work, as well as in the management of the university.

The viability of the Care Factor is being shown by Acumen, an EU Seventh Framework Program funded European project, which aims to find assessment parameters, not so much of the research as the work of researchers. For example, in its *Guidelines for Good Evaluation Practices* (April 2014), the calculation of academic age is based on a conventional value, which considers the number of children raised (p. 10), special allowances and other 'penalizing' factors (like illnesses, part-time jobs etc.).

The dream is that examples like Carol V. Robinson become much more common: Robinson went to work at age 16, then graduated and earned a bachelor's and a master's degree and finally, a PhD in chemistry. She then left the university for eight years to raise three children. Upon her return, she gained a professorship at Oxford on the basis of her research, becoming the first full female professor in chemistry. She also earned countless awards, including *Dame Commander of the Order of the British Empire.*

Robinson succeeded without the Care Factor. However, hers is also an isolated case. Can we do something to make her experience more widespread? The *Care Factor* helps us move forward in this direction.

References

Ashcroft, K., S. Bigger, and D. Coates (1996). *Researching into equal opportunities in colleges and universities.* London: Practical Research Series, Kogan Page.

Balbo, L. (1978). 'La doppia presenza'. *Inchiesta* 32: 3–11.

Barclay, L. and D. Lupton (1999). "The Experiences of New Fatherhood: A Socio-cultural Analysis". *Journal of Advanced Nursing* 29(4): 1013–1020.

Bentley, J. T. (2003). *Gender differences in the careers of academic scientists and engineers: A literature review.* Arlington, VA: National Science Foundation, Division of Science Resource Statistics.

Bezzi, C. (2001). *Il nuovo disegno della ricerca valutativa.* Milan: Franco Angeli.

Charyton C., J. Elliott, M. Rahman, J. Woodard, and S. Dedios (2011). "Gender and Science: Women Nobel Laureates". *Third Quarter* 45(3): 203–214

Chen, H. T. and P. H. Rossi (1981). "The multi-goal, theory-driven approach to evaluation. A model linking basic and applied social science". *Social Forces*, 59: 106–122.

Chen, H. T. (1990). *Theory driven Evaluation.* Thousand Oaks: Sage.

Chen, H. and P. Rossi (1989). "Issues in the theory-driven perspective". *Evaluation and program planning* 4: 299–306.

Cole, J. R. (1987). "Women in science". In: Jackson, D. N. and J. P. Rushton (eds) *Scientific excellence: Origins and Assessment.* University of Michigan: SAGE Publications: 359–375.

Committee on the status of women in sociology (2004). Washington, DC: American Sociological Association.

Corley, E., and M. Gaughan (2005). Scientists' participation in university research centers: What are the gender differences? *Journal of Technology Transfer* 30 (4), 371–381.

Corsi, M. and G. Zacchia (2013). "Che genere di economista: il possibile impatto delle nuove regole ANVUR". In *Roars*: <http://www.roars.it/online/che-genere-di-economista-il-possibile-impatto-delle-nuove-regole-anvur/>.

Dasaratha, V. R., K. Raghunandan, L. B. Logan, and B. V. Barkman (1997). "Gender differences in publications by promoted faculty". *Issues in Accounting Education* 12(2): 353–365.

Evans, E. and C. Grant (2008) (eds). *Mama, PhD. Women Write about otherhood and Academic Life*. New Brunswick, NJ: Rutgers University Press.

Franceschini, E. (2014), "Si lavora solo 11 ore la settimana". Sotto accusa mail e riunioni. Ecco tutti gli errori". *La Repubblica*, 4 December.

Gobo, G. (2008). *Doing Ethnography*. London: Sage.

Golde, C. M. and G. M. Walker (2006) (eds). *Envisioning the Future of Doctoral Education: Preparing Stewards of the Discipline. Carnegie Essays on the Doctorate*. San Francisco: Jossey-Bass.

Gornick, J. C. and M. K. Meyers (2008). "Creating gender egalitarian societies: An agenda for reform". *Politics and Society* 36(3): 313–349.

Harrington B., F. Van Densen, and B. Humbard (2011). *The New Dad: Caring, Committed and Conflicted*. Boston: Boston College Center.

Hochschild, A. (1975). "Inside the Clockwork of Male Careers". In: Howe, F. (ed.) *Women and the Power to Change*. New York: McGraw-Hill: 47–80.

Hochschild, A. and A. Machung (1990). *The Second Shift*. Avon Books: New York.

Hook, J. L. and C. M. Wolfe (2013). "Parental Involvement and Work Schedules: Time with Children in the United States, Germany, Norway and the United Kingdom". *European Sociological Review* 29(3): 411–425.

Jacobs, Jerry A., B. Keith, and H. A. Moore (1995). "Training Sociologists: An Assessment of Professional Socialization and the Emergence of Career Aspirations". *Teaching Sociology* 23 (3):199–214.

Jana, G. (2011). "Some sociological aspects of gender inequality in sharing housework in Slovene families and household". In: UNESCO (ed.) *Mapping the gender equality*: 173–189.

Kaufman, G. and E. Bernhardt (2014). "Gender, Work and Childbearing: Couple Analysis of Work Adjustments after the Transition to Parenthood". *Community, Work & Family*: 1–18.

Kennelly, I. and R. Spalter-Roth (2006). "Parents on the Job Market: Resources and Strategies That Help Academic Parents Attain Tenure-Track Jobs". *The American Sociologist* 37 (4): 29–49.

Knijn, T. and A. Smit (2009). "Investing, Facilitating, or Individualizing the Reconciliation of Work and Family Life: Three Paradigms and Ambivalent Policies". *Social Politics* 16 (4): 484–518.

Kyvik, S. (1990). "Motherhood and scientific productivity". *Social Studies of Science* 20 (1), 149–160.

Kyvik, S., and M. Teigen (1996). "Child care, research collaboration, and gender differences in scientific productivity". *Science, Technology and Human Values* 21 (1): 54–71.

Long, J. S., P. D. Allison P. D., and R. McGinnis, (1993). "Rank advancement in academic careers: Sex differences and the effects of productivity". *American Sociological Review* 58(5): 703–722.

Marotte, M. R., P. Reynolds, and R. Savarese (2010) (eds). *Papa, PhD. Essays on Fatherhood by Men in the Academy*. New Brunswick, NJ: Rutgers University Press.

Marra, M. (2012). "The Missing Links of the European Gender Mainstreaming Approach. Assessing Work-Family Reconciliation Policies in the Italian Mezzogiorno". *European Journal of Women's Studies* 19 (3): 349–370.

Marra, M. (2014). "Beyond Feminist and Gender-Based Approaches: Evaluating Cooperation between Men and Women for Gender Equity". In: Forss, K. and M. Marra (eds) *Speaking Justice to Power: Ethical and Methodological Challenges for Evaluators*. New Brunswick, NJ: Transaction Publishers: 89–111.

Marradi, A. (1981). "Misurazione e scale: qualche riflessione e una proposta". *Quaderni di Sociologia* 29(4): 595–639.

Marradi, A. (1990). "Classification, Typology, Taxonomy". *Quality and Quantity* 24 (2): 129–157.

Maske, K. L., G. C. Durden, and P. E. Gaynor (2003). "Determinants of scholarly productivity among male and female economists". *Economic Inquiry* 41(4): 555–564.

Mason, M. A. (May 3, 2012). "The Future of the PhD". *The Chronicle of Higher Education*. <http://chronicle.com/article/The-Future-of-the-PhD/131749>. Retrieved November 7, 2015.

Mason, M. A, N. H. Wolfinger and M. Goulden (2013). *Do Babies Matter? Gender and Family in the Ivory Tower*. New Brunswick, NJ: Rutgers University Press.

Mathews, A. L., and K. Andersen (2001). "A gender gap in publishing? Women's representation in edited political science books". *Political science and politics* 34 (1): 143–147.

Miller, T. (2011). *Making Sense of Fatherhood. Gender, Caring and Work*. Cambridge: Cambridge University Press.

NEF (2009). *A Bit Rich: Calculating the real value to society of different professions*. London: Report.

Palumbo, M. (2001). *Il processo di valutazione. Decidere, programmare, valutare*. Milan: Angeli.

Pawson, R. and N. Tilley (1997). *Realistic evaluation*. London: Sage.

Pescosolido, B. A. and R. Aminzade (1999) (eds). *The Social Worlds of Higher Education: Handbook for Teaching in a New Century*. Thousand Oaks, CA: Pine Forge Press.

Pillow, W. S. (2002). "Gender Matters: Feminist Research in Educational Evaluation". *New Directions for Evaluation* 96: 9–26.

Pizzorno, M.C., M. Benozzo, A. Fina, S. Sabato, and M. Scopesi (2014). "Parent-child career construction: A narrative study from a gender perspective". *Journal of Vocational Behavior*, 84: 420–430.

Rehel E. M. (2014). "When Dad Stays Home Too: Paternity Leave, Gender, and Parenting". *Gender & Society* 28(1): 110–132.

Robinson, D. (2006). *The status of higher education teaching personnel in Australia, Canada, New Zealand, the United Kingdom, and the United States*: Education International.

Rubery, J. (2002). "Gender Mainstreaming and Gender Equality in the EU: The Impact of the EU Employment Strategy". *Industrial Relations Journal* 33(5): 500–522.

Shalev, M. (2008). "Class Divisions among Women". *Politics and Society* 36(3): 421–444.

Sielbeck-Bowen, K. A., S. Brisolara, D. Seigart, C. Tischler, and E. Whitmore (2002). 'Exploring Feminist Evaluation: The Ground from Which We Rise'. *New Directions for Evaluation* 96: 3–8.

Spalter-Roth, R. and N. Van Vooren (2012). "Mothers in Pursuit of Ideal Academic Careers". *American Sociological Association, Department of Research and Development*: 1–12.

Stemwedel, J. D. (2009). "Engendered Species. Science, motherhood, and the Nobel Prize: have things gotten harder? Biochemical Evolution". *The Biochemical Magazine*, 31 (1): 62.

Suitor, J. J., D. Mecom and I. S. Feld (2001). "Gender, household labor, and scholarly productivity among University professors". *Gender Issues* 19 (4): 50–67.

Sweitzer, V. (2009). "Towards a Theory of Doctoral Student Professional Identity Development: A Developmental Networks Approach". *The Journal of Higher Education* 80 (1): 1–33.

Taino, D. (November 11, 2012). "Il buon lavoratore va a casa presto". *Corriere della Sera*.

Tower, G., Plummer, J. and B. Ridgewell. (2007). "A Multidisciplinary Study of Gender-based Research Productivity in the World's Best Journals". *Journal of Diversity Management* 2(4): 23–32.

Tucci, M. P., S. Fontani, and S. Ferrini (2010). "L'R-Factor: un nuovo modo di valutare l'attività di ricerca". *Studi e Note di Economia* 1: 103–140.

Rothausen-Vange, T. J., J. H. Marler, and P. M. Wright (2005). "Research productivity, gender, family, and tenure in organization science careers". *Sex Roles: A Journal of Research* 53(9/10): 727–738.

Van Dijk, T.A. (1977). *Text and Context,* London: Longman.

Verloo, M. (2005). "Mainstreaming Gender Equality in Europe. A Critical Frame Analysis Approach". *The Greek Review of Social Research* 117B: 11–34.

Wade, D. (2002). "Nobel Women". *Science* 295 (5554): 439.

Ward, K. and L. Wolf-Wendel (2012). *Academic Motherhood. How Faculty Manage Work and Family.* New Brunswick, NJ: Rutgers University Press.

Weiss, C. H. (1995). "Nothing as Practical as Good Theory: Exploring Theory-Based Evaluation for Comprehensive Community Initiatives for Children and Families'. In: Connell, J. P., A. C. Kubisch, L. B. Schorr, and C. H. Weiss (eds) *New Approaches to Evaluating Community Initiatives.* Volume 1. Concepts, Methods, and Contexts. Washington, DC: The Aspen Institute: 65–92.

Weiss, C. H. (1997). "Theory-based evaluation: past, present, and future". *New Directions for Evaluation* 76: 41–55.

Claudia Gualtieri

The Release of the Truth and Reconciliation Commission Findings on Indian Residential Schools in Canada, 2 June 2015[1]

> But for those thinking about what they could do now, here are three suggestions: (1) Read the TRC findings. [...] (2) Challenge colonial thinking. [...] (3) Listen to indigenous leaders, including survivors. [...] By listening and learning, by beginning to travel that road, we can work together to make the changes our society desperately needs.
>
> Jocelyn Thrope[2]

1. Introduction: Theory and Controversial Issues

The Truth and Reconciliation Commission (TRC) of Canada held its closing event in Ottawa from May 31–June 3, 2015. On June 2, a summary of the Commissioners' findings on Indian[3] residential schools was released in an official ceremony that was broadcasted in major cities across the country. In Ottawa, starting on May 31, the calendar of events was dense, meaningful, and moving. It included community actions of remembering, reconciliation, learning, and celebration, which reached its high point on June 3. This paper looks at this ground-breaking moment in contemporary Canadian history marked by the release of the TRC findings from the spatial location of Winnipeg. The province of Manitoba, with its large population of indigenous peoples, is indeed an advantageous location from which to observe a historical moment that directly addresses and involves

1 The author wishes to thank Maureen Matthews, curator of ethnology of the Manitoba Museum, for her useful comments and suggestions on this essay, and for her wonderful hospitality during the period of research in Winnipeg.

2 Thorpe, Jocelyn, "We can start to reconcile now", *Winnipeg Free Press*, June 16, 2015, p. 9.

3 The word "Indian" is employed here because it was the official name for residential schools in Canadian history. In other contexts, the terms "indigenous" and "aboriginal" will be used instead.

indigenous peoples. In addition, the TRC's headquarters are located in Winnipeg, and the research conducted by the commission and the statements gathered from the survivors will be stored at the National Research Centre for Truth and Reconciliation at the University of Manitoba. I was conducting research in Winnipeg and took part in the ceremony held on June 2, 2015[4].

The theoretical lenses through which the release of the TRC findings will be examined in this paper are borrowed from cultural studies and postcolonial theory, which means that this analysis will focus on two key notions: contextualization and colonialism. The historical, spatial, and cultural coordinates – the networks of relations, practices, ideas, and changes which shaped the context in which the examined event took place – are fundamental tools for exploring the dynamics that regulate institutional relations between Canada as a nation-state and the indigenous peoples who live within its borders. Likewise, the history of colonialism and its legacy – which still shape domestic politics – must remain in the foreground in order to understand the complex entanglement of processes and structures of power addressed by the release of the TRC findings.

To begin with, one should be reminded that although indigenous education has a long history in Canada, Indian residential schools initiated by Canada immediately after confederation were intended to facilitate the assimilation of indigenous peoples into colonial settler society.

This paper uses two case studies to analyze the ways in which the release was covered in two Canadian newspapers: *The Globe and Mail*, a national newspaper, and the *Winnipeg Free Press*, a local newspaper. The TRC website will also be examined and other institutional websites will be marginally referenced in order to shed light on the range of responses from the media, and to provide examples of how the official position of the Canadian government was called into question. From a cultural perspective, the relationships between indigenous and non-indigenous civil society and the federal government will be the main focus of the analysis. These networks of relations are analyzed in order to observe how these communicative networks help shape public discourses, along with political and cultural interactions in Canadian society.

Supported by a political rhetoric of emergency and fear, contemporary scenarios of global risk, economic crisis, and impending catastrophe have made utopian projects for better societies difficult both to conceive and plan. In this alarming context, the neoliberal paradigm has 'colonized' public imagination and

4 British Columbia is the original homeland of many indigenous Canadians who are actively involved in the process of reconciliation. See <http://reconciliationcanada.ca/>.

spread into all areas of civil society, supporting and justifying its own inevitability as the only viable path for addressing future challenges. In addition, theories of neo-managerialism in the public sector, and forms of new public management through information technologies have advanced the idea of a strategic pact between states and citizens along the lines of transparent communication. If these theories present reciprocal, profitable development both for the public sector and citizens, they indirectly offer a justification for the incessant renewal of the neoliberal system. From a critical standpoint, they undoubtedly carry the danger of reinforcing the hegemonic neoliberal economic agenda by suggesting a model of citizenship based on individualism, competition, and consumption. This non-inclusive prospective citizenship exposes ambiguous dynamics of inequality within the neoliberal paradigm. Connected to this, theories of new public management that confidently support the idealistic claim of successful communication as socially cohesive discoursive constructions, seem to overlook the important role that race and class play in society as practically joint categories of marginalization and exploitation.

In the case of indigenous peoples in Canada, these categories of subjection overtly surface in economic inequality, appearing in urban and land usage policies, different allocation of welfare services and education, and access to jobs. They still represent symbolic and practical boundaries within the economic neoliberal project applied to the public sector, which alarmingly risks severing the state from civil society. In this context, Larry Terry's warning in "Administrative Leadership, Neo-Managerialism, and the Public Management Movement" that "public entrepreneurs of neo-managerialist persuasion pose a threat to democratic governance" (Terry 1998: 194) is worth taking seriously. Indeed, the cases presented in this paper argue that public discourse and national management in postcolonial societies should address the traumas and divisions provoked by colonial history according to a long-sighted vision that extends beyond the Western neoliberal ideology, constructing entrepreneurial relations between states as companies and citizens as stakeholders.

The situation of Canada as a settler colony is rooted in the occupation of land inhabited by indigenous people by European colonizers. This historical fact still shapes issues of land property and rights, native titles, cultural recognition, definitions of identity and citizenship, access to national resources and welfare programs in both the national agenda and native claims. As postcolonial theory makes clear, the hybridization resulting from different cultural encounters and increased movement and migration complicates the ways in which concepts of identity, subjectivity and cultural affiliation may be constructed in contemporary societies, thus challenging the neo-managerialist, top-down administered structure of relations based on discursive techniques strategically organized as collective bonds.

In general, the work of Truth and Reconciliation Commissions as they operate in various contexts of conflict around the world is built on the significance of historical and personal memory, of truth and reconciliation, healing and forgiveness, silence and speech, and community and participation in both private and public domains. The ideal aim of TRCs is to attempt to re-read history in a way that might help increase cohesion and acceptance in society, hopefully opening up possible trajectories to modify governmental structures and institutional practices of subjection and exclusion. A critical view may suggest how the most impressive examples of TRCs (e.g. South Africa) have actually failed to substantially create a more equal and just society. In contrast, these cases often provide evidence of the most pervasive effects of economic globalization and increasing social injustice.

The TRC of Canada is an interesting example because it makes explicit how institutional impediments, political inconsistencies, and opposition towards social inclusion and sharing may obstruct and hinder the successful implementation of the Commission's work from the very moment these findings were released. Participative societies acting as national communities in postcolonial frameworks are difficult constructions whose utopian anticipation may be helped by processes of self-awareness, criticism, and shared responsibility encouraged through education, public debate, and political action, as both the Canadian TRC presentations and press coverage repeatedly emphasized.

One important educational step towards forging a communal sensibility among Canadian indigenous and non-indigenous peoples was taken by the Manitoba Museum in Winnipeg with the permanent exhibit "We are all Treaty People" that opened on August 12, 2015[5]. For the first time, all the Manitoba treaty medals were displayed along with the pipes and pipe bags used in the traditional ceremony. Maureen Matthews, the curator of ethnography at the Manitoba Museum and exhibit coordinator, explained that "[i]t is unusual for a museum to display sacred artifacts like these pipes and pipe bags, but without them we would have failed to represent First Nations agency and understandings"[6]. Matthews also emphasized the participation of indigenous peoples in the preparation of the exhibit: "We have collaborated with Treaty Relations Commission, consulted with the AMC Elders

5 <https://manitobamuseum.ca/main/museum-opens-two-new-permanent-exhibits/> (accessed September 28, 2015).

6 "Manitoba Museum makes aboriginal history permanent exhibit", *Winnipeg Sun*, 21 August 2015, <http://www.winnipegsun.com/2015/08/12/manitoba-museum-makes-aboriginal-history-permanent-exhibit> (accessed September 28, 2015).

Council and feasted the artefacts in advance of this exhibit"[7]. This innovative perspective voiced a wider need to re-examine the problematic history of treaties between British and Canadian colonial authorities and First Nations, indirectly addressing the history of the occupation of Manitoba by colonial Canada[8]. This latter issue is beyond the scope of this paper, though it is worth mentioning in order to raise a specific awareness of treaties as part of the process of re-reading, re-telling, and re-writing history from different perspectives, of which the colonial is but one. The master narrative of empire has dominated the historical record and its preservation in institutional archives, while indigenous views have largely been excluded and/or underestimated. Elders, indigenous lawyers and experts in Canada have worked through traditional and institutional channels in order to convince indigenous and non-indigenous peoples to adopt the idea, first voiced by Governor General Adrienne Clarkson, that "We are all Treaty people" and share treaties as part of a common identity, an outcome of a collective history now being re-formulated[9].

From these premises, it seems relevant to formulate crucial questions regarding the ways in which the process of historical reconstruction in postcolonial Canada may be read vis-à-vis scenarios of effective institutional communication to citizens. How was information about the TRC findings spread? Which key issues were focused on? Will the TRC findings released in June 2015, and the final report of December 2015, affect the agenda of the Canadian government after the general elections of October 19, 2015? Do discourses of neo-managerialism and nation branding affect the processes of healing and reconciliation heralded by the TRC? And, connected to this, may policies of multiculturalism help or hinder reconciliation[10]?

---

7 <http://www.communitynewscommons.org/our-city/education/we-are-all-treaty-people-exhibit-opens/> (accessed September 28, 2015).

8 See Pratt / Bone and the Treaty & Dakota Elders of Manitoba (2014) and D'Arcy / Bone and the Treaty & Dakota Elders of Manitoba (2014).

9 See: Treaty Relations Commission of Manitoba website, <http://www.trcm.ca/>; Office of the Treaty Commissioner website, http://www.otc.ca/; (), Government of Canada, Aboriginal Affairs and Northern Development Canada, A history of Treaty-Making in Canada, http://www.aadnc-aandc.gc.ca/eng/1314977704533/1314977734895;) and Government of Canada, Aboriginal Affairs and Northern Development Canada, Treaty Research Report, <http://www.aadnc-aandc.gc.ca/eng/1100100028653/1100100028654>.

10 In the past four decades in Canada, multiculturalism has been central to the national agenda, particularly when applied to the issue of migration. Provisions related to multiculturalism were included in the Constitution Act of 1982. Canada's Multiculturalism

It is important to define ways in which the connection between neo-managerialism, nation branding, and multiculturalism will be applied here. This paper will not consider how the cultural wealth of the nation is exploited for tourism and nation branding[11], or how native cultures and their symbolic resources may be employed as modes of development of national economies[12]. The relationship between nation branding and neo-managerialism will be used to problematize ways in which scenarios of democratic participation may be thought of, and pursued in postcolonial contexts, thus affecting the dynamics and politics of the inclusion of cultural diversity in society. Since neo-managerialism and nation branding share a common concern of employing effective communication as a means for social cohesion, their association with multiculturalism may help to expose the inherent ambiguities that lie beneath and complicate their practical use as political projects of social participation. This will hopefully provide insight into the ways multicultural policies may be thoughtfully and fruitfully implemented. In a provocative reading of Canadian multicultural policy in *Multiculturalism and the History of Canadian Diversity*, Richard Day interrogates what he calls "the problem of the problem of diversity" throughout Canadian history, and challenges the notion of multiculturalism as a Canadian obsession[13]. He argues that no state intervention can bring an end to tensions related to ethno-cultural relations of power, and suggests that the idea of unity imposed by the nation-state must be abandoned in order to pursue effective politics of multiculturalism. According to this perspective, the very notions of diversity and unity are questioned and set up as fields of contention within culture and society, which also underscores the demand for reconstructing their meanings within the frame of the nation-state.

Richard Day's argument can be put together with Melissa Aronczyk's claim in *Branding the Nation* to highlight the links between managing and branding the

---

Act was adopted in 1988. The ideological and political experiment of multiculturalism marked an attempt to frame diversity as unity, which was consistent with the constructed mosaic image of the nation. There is a vast literature on Canadian multiculturalism that features both harsh criticism and defensive assessment of this national policy, also comparing the Canadian form of multiculturalism to similar political projects in different countries. See: Banting and Kymlicka (2010); Kymlicka (2012); Garcea, Kirova and Wong (2008); Banting, Courchene and Seidle (2007).

11 See Bandelj and Wherry (2011).

12 In colonial imagery, indigenous cultural expressions and artefacts are often perceived as exotic stereotypes that match the romantic representation of the primitive good native or of the savage Indian in Hollywood Western films.

13 See Day (2000: 3).

nation. Aronczyk advances the idea that "nation branding has become a solution to perceived contemporary problems affecting the space of the nation-state: problems of economic development, democratic communication, and especially national visibility and legitimacy amid the multiple global flows of late modernity". Because nations have lost their ideological and cohesive allure as imagined communities (to borrow Benedict Anderson's famous definition) in the era of economic globalization and domineering finance, they have gradually transferred their appeal to quality branding in order to attract world capital into national economies and to further reassert the driving force of economic neoliberalism. By adopting this view, managing the nation according to economic neo-liberalism and, at the same time, pursuing ideals of democratic citizenship seems to be an unconvincing argument. This is because the management of public resources and discourses according to technocratic self-empowering economic principles works against democratic and representational forms of government that serve the common good, which stem from and are endorsed by heterogeneous civil society. As Aronczyk argues, "[n]ation branding represents a transformation of business in the articulation of national identity. [It] reveals how the social, political, and cultural discourse constitutive of the nation has been harnessed in new ways, with important consequences for both our concept of the nation and our ideals of national citizenship" (Aronczyk 2013: 3–4). This reading offers a useful background against which an analysis of the release of the TRC findings may be contextualized, and sheds light on how complex and multifaceted the case appears if one adopts an indigenous perspective. In fact, as the work of the TRC highlighted during the years of hearings, the place and presence of First Nations, Inuit, and Métis peoples in Canadian contemporary history, politics, and society have yet to be affirmed with respect to their traditional histories and beliefs.

It is therefore useful to analyze the TRC release and the summary of its final report to illustrate how the scars and shadows of colonialism survive and remain active and present in contemporary Canada. Canadian colonialism produced a radical disparity of power, created forced and dysfunctional relationships, and brought about cultural alienation typified by stereotypes, institutional racism, and inconsistencies of bureaucracy. In fact, while governmental agencies over the years have produced data on their own failures with regard to indigenous issues, remedies have been sought in civil society in an attempt to accommodate the rights and recognition owed to indigenous peoples[14]. Education is emphasized by

14 A reassuring and inspiring analysis of the ways in which "[t]he promise of democracy rests on the practice of active citizenship" is offered by David Campbell in the

the TRC as paramount in raising awareness about different historical narratives. Within this history, past and present colonial practices should be remembered, revealed, investigated, and evaluated in order to inaugurate a symbolic process of healing and reconciliation.

2. The Colonial History of Indian Residential Schools in Canada[15]

Indian residential schools were boarding schools for First Nations, Inuit[16], and Métis[17] children funded by the Canadian government and mainly run by administrative and religious institutions. Pre-confederation cuts of the British Parliament in the mid-nineteenth century anticipated the system eventually created[18]. In 1844, the Bagot Commission recommended making use of indigenous education as a means of assimilation. According to the 1867 British North America Act, powers having to do with the education of Canadians were allocated to provincial governments, while the treaty for Indian people's education fell under the jurisdiction of the federal government[19]. In 1879, Nicholas Flood Davin's *Report on Industrial Schools for Indians and Half-Breeds*, commissioned by Prime Minister John A.

---

fascinating 2015 essay *Metis, craft, civic mindedness: Essential attributes of democratic citizenship in communities.*

15 See Government of Canada, Aboriginal Affairs and Northern Development Canada, Indian Residential Schools, <https://www.aadnc-aandc.gc.ca/eng/1100100015576/1100100015577> and Indian Residential Schools Resources website, http://irsr.ca/.

16 The involvement of Inuit children in residential schools differs from that of First Nations children. The Inuit population was affected by the Indian Act of 1924 and constitutionally classified as "Indian" in 1939. Schools for Inuit children began to operate in the North in the mid 1950s. <http://www.aadnc-aandc.gc.ca/eng/1100100016900#chp2> (accessed September 30, 2015).

17 It was with the Northwest Half-breed Claims Royal Commission of 1885 that the federal government addressed the issue of Métis education. http://www.aadnc-aandc.gc.ca/eng/1100100028809 (accessed September 30, 2015). Ways in which residential schools affected Métis children and culture, and the Métis community's involvement in the telling of the history of residential schools, are often perceived as being overlooked and underestimated.

18 See Government of Canada, Aboriginal Affairs and Northern Development Canada, Historical Legislation, <https://www.aadnc-aandc.gc.ca/eng/1100100010193/1100100010194>.

19 Part of the legislation about assimilation was the Gradual Civilization Act of 1857, the Act for the Gradual Enfranchisement of Indians of 1869, and the Indian Act of 1876.

MacDonald, provided the rationale for the public funding of residential schools[20]. Subsequent legislation made residential schools the standard educational system for these indigenous children, and attendance was made compulsory both for day schools and residential schools, depending on availability in the territory. Historical records are inconsistent when it comes to locating and naming schools. It seems that the first ones were established in the 1830s, and by 1931, a total of 80 active schools were reported. The last residential school run by the government was closed in 1996. Approximately 132 residential schools operated between 1831 and 1996. For the purpose of providing compensation to former students, the Indian Residential School Settlement Agreement identified 139 residential schools[21].

Residential schools were part of a structured and strategically devised system aiming to assimilate indigenous peoples into the Euro-Canadian colonial society[22]. Ideologically colonial institutions, residential schools helped to justify European colonization and Christianization; their civilizing mission was perpetrated on people perceived to belong to a morally inferior race. It was part of the British colonial policy to try and eradicate indigenous cultures from conquered lands, and to physically eliminate traces of cultural indigeneity in the native people[23].

A similar project was undertaken with the so-called half-blood indigenous peoples of Australia, which was, like Canada, a British settler colony. In Australia, "stolen generations" were part of a planned systematic program of action to irreversibly separate children from their families and eliminate the Aborigines from the newly settled land. Assimilation also aimed to introduce indigenous children into colonial society as a new labor force. To this end, they were trained in manual and subservient work. By the mid-nineteenth century in Canada, farming had become the economic engine of the prairie, replacing the centrality of the fur trade of the previous 200 years. On the plains, agriculture became a basic productive activity in the colonial settler economy, which disrupted indigenous systems of

---

20 The document is conventionally called "The Davin Report".

21 www.trc.ca.

22 In 1920, Duncan Campbell Scott, Superintendent of Indian Affairs from 1913 until 1932 made the government's position towards native peoples explicit by stating: "I want to get rid of the Indian problem […]. Our object is to continue until there is not a single Indian in Canada that has not been absorbed into the body politics, and there is no Indian question, and no Indian Department" (*Report of the Special Committee of the House of Commons examining the Indian Act amendments of 1920*).

23 See *Still taking 'Indian' out of the child?* (*Winnipeg Free Press*, 2015: 8). This colonial strategy of assimilation, recalls, in its most abjected form, the explicit sentence "Exterminate all the brutes" addressed to Africans in Joseph Conrad's *Heart of Darkness*.

land use, making the need for new territory a central obsession of colonial occupation. Instructing pupils to work the land at residential schools reinforced this strategy of assimilation.

Historical records and the TRC findings document the complex variety of situations and conditions under which children were sent to residential schools. However, even if it is debatable which pressures and constraints individual families endured, what is unmistakably evident is the overall callous treatment and harsh conditions endured by most of the children. The system was meant to isolate children from their culture of origin: in many schools they were forbidden to speak their languages or practice their beliefs, were kept away from their families for long periods, and rebuked for transgressing the rules. This separation from their family and original community and culture was coupled with instances of cruel punishment, despicable violence, and physical and psychological abuse. Furthermore, schools were overcrowded and inadequately funded. Poor sanitation and lack of medical care caused disease and death in high numbers[24]. The TRC found that about 150,000 indigenous children attended residential schools, and more than 1,000 died at these institutions. The exact number is not known because no official death records were kept, and many children were buried in unmarked graves.

The legacy and long-lasting effects of residential schools on Canadian indigenous peoples generated an unresolved historical trauma that has been described as "cultural genocide"[25]. The conditions of indigenous peoples in today's Canadian society bear testimony to this trauma as a collective wound, seen in the high rates of criminal convictions, alcoholism, drug addiction, lack of education and skills, psychological disorders, and low income[26] among indigenous Canadians. In addition to these psycho-social factors, inter-generational suffering and the violent consequences of family dysfunction are passed on, thus reproducing historic trauma across the generations[27]. First and second-generation survivors of residential schools are now pulling together their efforts, calling upon the Canadian government and society to share this painful historical legacy. Memory and healing are key words in this process of sharing and reconciliation. Their symbolic power provides a guiding path in order to identify ways in which justice and reparations may begin to take effect.

A Truth and Reconciliation Commission was active in South Africa from 1995 until 1998, facilitating the peaceful transition from apartheid to a post-apartheid

---

24 See Bryce (1922).

25 See Pankratz (2015: 9).

26 See *From residential schools to jails* (*Winnipeg Free Press*, 2015: 8).

27 See Reconciliation, <www.trc.ca> (accessed September 30, 2015).

democracy. Through voluntary confessions, both victims and perpetrators of racial crimes would be publicly forgiven, provided they spoke the truth. Victims could have access to reparations and compensation, while amnesty was guaranteed to perpetrators. By comparison, the Canadian TRC did not mark a radical change in Canadian history and politics, and had no official or legal power to support the witnesses or to offer amnesty in exchange for truthful accounts from the staff of these residential schools or to those identified as having abused indigenous children. Instead, the Canadian TRC mainly acted to record the accounts of victims and listen carefully and sensitively to their testimonies. Little evidence was provided by the staff of the schools and by representatives of official institutions[28]. Financial compensation for victims was outside of the scope of the TRC, but survivors of residential schools were provided with compensation by a parallel process that was co-funded by the churches that ran the schools and the Government that funded them.

While symbolic reparations may yet be pursued through networks of solidarity and respect for human rights, financial compensation and long-term projects aimed at transforming the poor conditions of indigenous peoples in Canadian society and at reshaping the relationship between indigenous and non-indigenous Canadians, progress will not happen unless positive action is included in the political agenda. Contrasting the TRC process with the South African model, the editorial published on June 1, 2015 in the *Winnipeg Free Press* noted the absence of the Canadian state: "Unbelievable that with all this time and attention spent on TRC, the federal government can't be counted on to show true leadership". Civil society is said to have taken the lead, "[e]ven the chairman of the TRC, Murray Sinclair, is not expecting much from Mr. Harper. Instead, he's hoping that charities and non-profit groups, schools and educators, academics and neighborhood associations will engage to discuss the findings"[29].

3. The Truth and Reconciliation Commission on Indian Residential Schools[30]

First Nation activists and leaders brought the issue of residential schools to the fore of national politics, and investigations were triggered by institutional agencies, which involved enquiries into the status of native peoples in Canada. A Royal

28 See Galloway (2015: 1, 6–7).

29 See *Still taking 'Indian' out of the child?* (*Winnipeg Free Press* 2015: 8).

30 Full text is available at <http://www.parl.gc.ca/Content/LOP/researchpublications/prb0848-e.pdf> (accessed October 5, 2015).

Commission on Aboriginal Peoples (RCAP), established in 1991, issued a report in 1996 examining historical relations and legal agreements since colonial times, and provided the Canadian government with detailed indications for implementing changes[31]. In 1998, the federal document "Gathering Strength: Canada's Aboriginal Action Plan" openly addressed Indian residential schools. It contained a statement of reconciliation and a plan for political action, which was implemented with a compensation package. The Aboriginal Healing Foundation (AHF) was established to support community-based actions for former students. In 2000, the Aboriginal Healing Foundation founded the Legacy of Hope Foundation in order to raise awareness about residential schools and to support survivors.

The Indian Residential Schools Settlement Agreement (IRSSA), which was established in 2007 as a result of a major class action settlement in Canadian history, began to allocate financial compensation according to specific criteria for the damage inflicted to First Nations, Inuit, and Métis students in residential schools – a process which continues today[32]. These criteria relate to the time and place of attendance at residential schools, forced attendance, subjection to serious physical and sexual abuses, and the quality of testimony[33]. Proving eligibility may present a twofold problem because schools were not always properly registered and documentation of students in attendance was often incomplete. In addition, testimonies were provided orally. Oral evidence is often thought to be secondary to written documentation and reminds us of the multiplicity of ways of interpreting, narrating, and archiving history in cultures where orality is a trusted source of knowledge and a reliable basis for the transmission of history. The Truth and Reconciliation Commission on Indian Residential Schools in Canada worked to keep this indigenous principle in mind.

The duty of the commissioners of the TRC was to gather, document, and preserve the survivors' memories. The Commission was launched on June 2, 2008. After initial obstacles, the Truth and Reconciliation Commission was established

31 Full text available at <http://www.parl.gc.ca/Content/LOP/ResearchPublications/prb9924-e.htm>; see also <http://www.aadnc-aandc.gc.ca/eng/1307458586498/1307458751962>; and <https://archive.org/details/RoyalCommissionOnAboriginalPeoples-FinalReport-Vol.1-Looking>, (accessed October 5, 2015).

32 The IRSSA financed the Independent Assessment Process (IAP), which was mainly concerned with claims of serious physical and sexual abuse. Hearings are expected to continue until around 2017. See <https://www.tbs-sct.gc.ca/rpp/2008-2009/inst/ira/ira02-eng.asp>.

33 Competent bodies include the Common Experience Payment (CEP), Independent Assessment Process (IAP), and Truth and Reconciliation Commission (TRC).

in October 2009 by Governor-General Michaëlle Jean. It was chaired by Justice Murray Sinclair, an Ojibwe-Canadian judge, with Commissioners Chief Wilton Littlechild, former Conservative Member of Parliament and Alberta Regional Chief for the Assembly of First Nations, and Dr. Marie Wilson, a senior executive with the Workers' Safety and Compensation Commission of the Northwest Territories and Nunavut. The Truth and Reconciliation Commission was assisted by an Indian Residential School Survivor Committee (IRSSC) composed of 10 representatives from various Aboriginal organizations and survivor groups selected by the Federal Government in consultation with the Assembly of First Nations (AFN). An Inuit sub-commission was also established, headquartered in Yellowknife, in the Northwest Territories. It ensured that Inuit survivors of the residential schools were included in the national truth-telling and reconciliation process. From January 2011 to April 2012, the AFN carried out work on behalf of the TRC in Nunavut, the Yukon and the Northwest Territories under the co-direction of Jennifer Hunt-Poitras and Robert Watt[34].

The TRC mandate was originally scheduled to complete its report in five years[35]. However, it was extended to 2015 due to the number of records provided by survivors. The Commission explored activities alleged to have occurred in residential schools and their negative impact. As the TRC website states, most of the statements were gathered at national and regional events, and at community hearings. Some were presented publicly in panels and circles[36]. Seven national events were held during 2008–2013. The investigation brought to light appalling events related to the students' deaths at these institutions, including burials in unmarked graves without the parents' notification or consent[37]. To supplement and confirm this information, the Commission gained access, through a court order, to historical

34 See <http://www.trc.ca/websites/trcinstitution/index.php?p=158> (accessed October 15, 2015).

35 Full text is available at <http://www.trc.ca/websites/trcinstitution/File/pdfs/SCHEDULE_N_EN.pdf> (accessed October 15, 2015).

36 "Share your truth", www.trc.ca (accessed September 30, 2015).

37 As *The Guardian* reports in an interview to residential school survivor Sue Caribou, "As many as 6,000 children died in residential institutions, which ran from 1876 to 1996. The accurate figure could be much higher however, since the government stopped recording aboriginal students' deaths in 1920 in light of the alarming statistics. Caribou believes that dozens of pupils perished at the institution where she was detained. 'Remains were found all over the fields. But numbers do not reflect the reality. Many of my friends committed suicide after their release', said Caribou, who said she was frustrated that an inquiry did not take place twenty years ago, after the last of the residential schools closed." "Canada confronts its dark history of abuse in residential

records in the National Archives during 2012–2013. As Adele Perry writes in an article from the *Winnipeg Free Press* on July 10, 2015, "[t]his history is not over": the effect of the TRC's work has just begun. Its fundamental achievement being shared responsibility and common action for reconciliation and change. "'Reconciliation is not an aboriginal problem,' said Sinclair. 'It is a Canadian problem'"[38].

By June 2, 2015 the TRC had compiled a summary titled 'Honouring the Truth, Reconciling the Future', which anticipates the six-volume Final Report, published in December 2015. The summary contains 94 recommendations, namely, "Calls to Actions", grouped in the categories of legacy and reconciliation, which precisely identifies major fields for future interventions. "Calls to Action" aims at rebuilding and radically modifying the relationship between indigenous and non-indigenous Canadians. In an interview, Justice Murray Sinclair related reconciliation to education, and expanded the conventional meaning of education to include adult people at all levels of society. He stated that "[p]art of the misunderstanding that we see so prevalent in Canadian society is young adults, and adults in positions of leadership, constantly demonstrate a total lack of understanding and misunderstanding about who aboriginal people are"[39]. Here, he launched a clear attack against persistent colonial attitudes towards indigenous peoples in Canadian society, thus laying the foundations of the national debate on the recognition of the colonial ideological legacy.

4. Media Coverage of the Release of the Truth and Reconciliation Commission findings

A chronological reading of the media coverage of the release of the summary of the TRC findings began on May 29, 2015, when, in a speech given at the fourth annual Pluralism Lecture of the Global Centre for Pluralism in Ottawa, Supreme Court Chief Justice Beverly McLachlin set the tone of the political and cultural debate by explicitly maintaining that Canada's residential schools policy was an

schools," <http://www.theguardian.com/world/2015/jun/06/canada-dark-of-history-residential-schools> (accessed 30 September 2015).

38 Robson, Mia, "Canadians United for change", *Winnipeg Free Press*, June 1, 2015, p. 3.

39 Chiose, Simona, "Universities must address residential schools legacy, Justice Sinclair said", *The Globe and Mail*, May 29, 2015, <http://www.theglobeandmail.com/news/national/education/universities-must-address-residential-school-legacy-justice-sinclair-says/article24709958/> (accessed October 5, 2015).

attempt to commit "cultural genocide" against indigenous peoples[40]. The term "cultural genocide" was echoed in many newspaper articles, building on the existing debate about the extent to which residential schools had contributed to the disruption and destruction of indigenous cultures in Canada[41]. The emphasis on "genocide" and "cultural genocide" conjured an immediate connection to the Jewish holocaust, addressing complex legal matters[42], and calling into question Canada's official rejection of the United Nations Declaration on the Rights of Indigenous Peoples[43].

Experts weighed the UN definition of "genocide" and its possible applications against the system of forced assimilation in Canada. Although the United Nations Convention on the Prevention and Punishment of the Crime of Genocide, adopted in 1948, did not use the phrase "cultural genocide"[44], the debate showed that "[i]t is broader than just an aboriginal issue"[45]. Experts also tried to map out the political effects of McLachlin's statement, even if, as many believe, it would not practically change current Canadian legislation. Nevertheless, one of the recommendations in the TRC "Calls for Action" is exactly that Canada reconsider its position on the United Nations Declaration on the Rights of Indigenous Peoples. In an article entitled 'McLachlin said what many have long known', scholar and treaties expert Ken Coates emphasizes how much words matter, arguing that it is

---

40 Fine, Sean, "McLachlin: A history of 'cultural genocide'. Supreme Court Chief Justice delivers 'unparalleled' remarks over Canada's treatment of First Nations people", *The Globe and Mail*, June 29, 2015, pp. 1, 12.

41 Chief Justice McLachlin authored the important decision about indigenous rights of June 2014 to grant title to the Tsilhqot'in nation. It determined the terms of ownership of ancestral land for indigenous people. See Kopecky (2015: 30–39).

42 See Fine (2015: 6).

43 Full text available at <http://www.un.org/esa/socdev/unpfii/documents/DRIPS_en.pdf> (accessed October 15, 2015). See also Towtongie and Coon Come (2015). The article severely criticizes the federal government's opposition to the implementation of the United Nations Declaration: "Ultimately, what lies behind the rhetoric about the declaration is the federal government's resistance to sharing, constructively and creatively, the extraordinary power and control it exercises over so many indigenous peoples in Canada, notwithstanding it was the misuse of this power that impoverished and dispossessed indigenous peoples. That is the real heart of the matter."

44 Full text available at <https://treaties.un.org/doc/Publication/UNTS/Volume%2078/volume-78-I-1021-English.pdf> (accessed October 5, 2015).

45 These are the Northwest Territories Minister of Education and former residential school survivor Jackson Lafferty's words, as they are reported in Galloway and Curry (2015: 10).

important to change both the national vocabulary and the ways in which official history has been constructed. He contests the old narrative of Canadian benevolence and argues in favor of a more realistic story based on the recognition of injustice and on the acceptance of shared responsibilities. However, in spite of the media coverage of the TRC findings in both the national newspaper *The Globe and Mail* and the *Winnipeg Free Press*, and notwithstanding what was perceived to be the popular support for accepting the term "cultural genocide"[46], statistics showed that few Canadians actually paid attention to the issue of residential schools[47]. This poses serious questions as to what extent the official discourse of the nation – defined as addressing model citizens in national official and governmental communication – can be expected to engage with the indigenous question, and to speak to indigenous peoples as citizens.

In 2008, Prime Minister Stephen Harper apologized to indigenous peoples for the residential schools system and its program of assimilation, but his government denied that it was a form of genocide, calling it instead "forced assimilation"[48]. Since then, apologies seem inconsistent and there is still a widespread cynical attitude amongst indigenous people and their allies as to whether the federal government is going to take any action based on the TRC's suggestions[49]. In contrast, immediate political action seemed to be understood by the TRC, activists, and journalists, as the only possible path in order to set up a line of continuity with past positive efforts and institutional progress. Actions were intended in diverse fields of application.

The TRC argued that improving indigenous education and supporting indigenous cultural preservation was their foremost recommendation, joined with a strong emphasis on the revision of colonial history, and a concrete actualization of

---

46 Rabson (2015: 12).

47 Rabson (2015), <http://www.winnipegfreepress.com/canada/New-poll-shows-many-Canadians-not-paying-attention-to-issue-of-residential-schools-312964481.html> (accessed October 1, 2015).

48 "'Canada confronts its dark history of abuse in residential schools': Landmark report reveals school system's brutal attempt to assimilate thousands of native children for more than a century and gives voice to survivors", <http://www.theguardian.com/world/2015/jun/06/canada-dark-of-history-residential-schools> (accessed September 30, 2015). In 2009, Pope Benedict XIV released an official statement about the Roman Catholic Church's role in residential schools.

49 See Lett (2015: 5): "Harper repeatedly referred to his 2008 apology and to ongoing government programs to help improve conditions for aboriginal Canadians. The political translation: We have done everything that we're prepared to do."

legal justice for indigenous peoples. "There is no forgiveness without recognition"[50], as Murray Sinclair stated, and recognition is achievable through education and the application of equal justice. In an article entitled 'Education a way to reconciliation: justice', Simona Chiose presented interviews with indigenous leaders and law experts highlighting the importance of popular actions, such as the Idle No More movement, mainly led by indigenous women in the cities[51]. Movements of people were perceived as a powerful means for raising a collective voice in favor of historical and cultural recognition. Risks were underscored, such as "the difficulty of making space for difference without romanticizing Aboriginal identity", the huge contrast between "Western legal concepts [...] rooted in concepts of property and individualism [...] and indigenous legal traditions [...] that ideal of living a good life and a collective life"[52], and that "the issue has become too 'compartmentalized', too detached from other injustices embedded in Canadian history"[53]. Educational institutions were cited as proper places where these dangers might be confronted and controlled. Suggestions proposed that more indigenous material be included in post-secondary courses, that universities try to capture multiple cultural viewpoints and that revised versions of colonial history be told in schools nationwide. Finally, the history of residential schools should become part of the educational curriculum in every province[54].

According to the TRC's recommendations in 'Honouring the Truth, Reconciling for the Future', change in educational practices rests on a radical revision of colonial history, which will also bring about important modifications in the national legal system. At present, it is considered essential that Canada officially reject both the Doctrine of Discovery and the notion that the lands claimed by European conquerors were *terra nullius*[55]. As ideological positions and legal principles, both notions helped to justify and consolidate colonial occupation. They still impinge on national culture, administrative practices, and political choices, thus preventing a successful process of reconciliation and the realization of a just society.

This pressing drive towards a different narrative of Canadian national history triggered debate among scholars, teachers, journalists, and representatives of

50 See Chiose (2015: 5).
51 <http://www.idlenomore.ca/>
52 See Chiose (2015: 5).
53 See Swan (2015: 3).
54 Galloway and Curry (2015: 1).
55 The Truth and Reconciliation Commission of Canada, *Honouring the Truth, Reconciling for the Future. Summary of the final report of the Truth and Reconciliation Commission of Canada*, 2015, www.trc.ca (accessed October 1, 2015).

various institutions. "Are we up to the challenge of rebooting history?", wondered Dan Lett, analyzing how television programs still perpetuate stereotypes brought to Canada by European settlers. Lett also emphasized how insidious elements and incorrect ideological assumptions in the national narrative of history need to be deconstructed in order to reformulate it in more fair and balanced terms[56].

Conflicting ways of interpreting history and enacting change were evident from the very day the TRC findings were released. In response to Jeffrey Simpson's 'Fixating on the past is not productive' which alerted readers to the danger that "a relentless fixation on 'the past', especially the tragic and embittered parts, risks driving elements of a society apart, making reconciliation paradoxically more, not less, difficult"[57], historian Adele Perry stresses ways in which "[h]istory is fundamentally about interpretation. [And] the TRC Summary does demand that we place indigenous peoples' version of history of residential schooling, and indeed Canada, at the center rather than the margins of the history we tell"[58]. Opposing versions of history provoked indignant reactions. In an article titled 'Debunking the half-truths and exaggerations in the Truth and Reconciliation report' published on the National Post website on June 4, 2015, former University of Manitoba professors Rodney Clifton and Hymie Rubenstein[59] expressed their dissatisfaction with the TRC report for having exaggerated the negative impact of residential schools and for providing incomplete evidence of positive examples of formal education in these institutions[60]. In reply, Norm Gould referred to these remarks as "retrograde opinions". He argued against this form of historical denial and proposed that academic freedom and responsibility were indispensable tools to overcome this phenomenon[61]. In a similar vein, Dan Lett replied with a scathing article in response to Bill Marantz's online statement on Winnipeg's *Jewish Post and News*, in which he referred to the TRC report as "the Half Truth and Recrimination Report"[62]. Milder criticism of the TRC's work included suggestions

56 See Lett (2015: 4).
57 See Simpson (2015: 11).
58 See Perry (2015: 9).
59 See Geary (2015: 13).
60 See Clifton and Rubenstein (2015) <http://news.nationalpost.com/full-comment/clifton-rubenstein-debunking-the-half-truths-and-exaggerations-in-the-truth-and-reconciliation-report> (accessed October 10, 2015).
61 See Gould (2015: 9).
62 See Lett (2015) <http://www.winnipegfreepress.com/opinion/columnists/Apology-welcome-but-doesnt-excuse-condemned-editorial--311029771.html> (accessed October 30, 2015).

that the Commission might have gone further in investigating events and, most of all, in specifying decisive solutions[63].

While it was clear that the media and public debates were active, and hopefully useful for future development, contrasting factors emerged concerning the scope of the discussion, on the one side, and the silence of federal institutions, on the other. The complex network of ideas onto which practices of assimilation, cultural and physical genocide are entangled in history show how any process of reconciliation must be rooted in historical recognition. As Adam Muller argues in 'Indian residential schools were unqualified genocide', many "have drawn attention away from myriad, long-standing, and enduring ways in which unqualified genocide has been perpetrated in Canada. [...] To admit less than this [...] is to blur the distinction between the methods and the aims of genocidal conduct"[64].

In Canada, while investigations and reflections dug deep into histories of cultural destruction, multiple institutional strategies of subjection surfaced, and dangers of marginalization emerged. The discussion showed the connection between the residential schools system and the "Sixties Scoop", a child welfare program that took indigenous children away from their parents and adopted them into non-indigenous families, mainly in the United States. The survivors are now asking for similar justice and reparation[65]. The Métis people, too, who have a complex and specific group history in Canada, claim that they were ignored in the process of historical reconciliation led by the TRC[66]. In addition, the contemporary emergence of the issue of more than one thousand missing and murdered indigenous women illustrates that the problem of indigenous children's welfare is still not over, which further complicates the indigenous question in Canadian society. These contemporary issues are also in desperate need of prompt institutional action[67].

In spite of positions of active engagements proposed by local institutions voiced in the press and public channels, the federal government – which bears the legal responsibility for the victims of residential schools – has remained silent. Speaking on behalf of Canada's provincial and territorial leaders, Paul Davis,

---

63 See Perry (2015: 9): "In some ways, the TRC summary report doesn't ask for enough"; 'Join them on a path to healing' (Editorial, *Winnipeg Free Press*, 2015: 6).

64 See Muller (2015: 9).

65 See Puxley (2015: 9).

66 Conversation between Claudia Gualtieri and Lawrence Barkwell at the Manitoba Métis Federation in Winnipeg, June 15, 2015, not published; Puxley, Chinta, "Métis fear they are being sidelined", *Winnipeg Free Press*, June 16, 2015, p. 4.

67 See <http://www.theguardian.com/world/2015/jun/06/canada-dark-of-history-residential-schools> (accessed October 25, 2015).

Newfoundland and Labrador Premier, declared that "Canada's premiers support the Truth and Reconciliation Commission's recommendations and will act on them with or without Ottawa's help". He also emphasized that "Prime Minister Stephen Harper's longstanding absence from first ministers' meetings is a missed chance for collaboration"[68]. In addition, responding to the TRC's call for action, an agreement to develop a strategy to preserve and promote indigenous languages was signed by the provincial government of Manitoba together with indigenous and local educational and cultural agencies. However, as Nick Martin ironically remarked, "there was one potentially useful partner missing Monday — the federal government"[69]. The federal government will provide some financial support, Martin added, but it has remained an external, peripheral, and uninvolved actor[70].

At the formal ceremony of the release of the Summary of the TRC findings in Ottawa on June 2, 2015, the federal government's silence and passivity were brought to the forefront by media analysts. Bill Curry noted that "Stephen Harper took part in an emotional closing ceremony for the Truth and Reconciliation Commission, but the Prime Minister did not say a word"[71]. And Shannon Sampert, one of the few white pupils in the residential schools system, stated: "I would have liked to see my prime minister and his ministers sitting in the audience in Ottawa on Tuesday, surrounded by survivors. [...] In an era of optic politics, it would have sent a clear message this is being taken seriously, or that their voices have been heard"[72]. Sampert also emphasized that the State was highly underrepresented at the official release. And Phil Fontaine, former National Chief of the Assembly of First Nations, defined the government's prolonged silence as "confusing"[73].

The Canadian Prime Minister did deliver a written speech after the release of the Summary of the TRC findings, stating that: "While this is an important milestone in getting our country past the days of Indian residential schools, work is still

---

68 See Bailey (2015) <http://www.winnipegfreepress.com/canada/native-leaders-to-discuss--violence-children-in-care-at-meeting-with-premiers-315246151.html> (accessed October 1, 2015).

69 See Martin (2015) <http://www.winnipegfreepress.com/local/Mutlitude-of-groups-signs-on-to-preserve-foster-indigenous-languages-328513801.html> (accessed October 1, 2015)

70 See Curry (2015: 1, 9).

71 See Curry (2015: 1, 9).

72 See Sampert (2015: 9).

73 See Kusch (2015) (accessed October 1, 2015).

needed to help heal the pain and restore trust from that wrong"[74]. The government spoke in favor of informed and well-reasoned decisions, as the official website briefly stated[75]. The general feeling was that the government was taking its time, waiting for the federal elections of October 19, 2015 and the release of the final TRC report to take place. However, the article 'Truth and consequence' published in *The Economist* on July 6, 2015, considered how hopes might be frustrated, given that the expectations raised by the Prime Minister's apology in 2008 were left without effects, in the same way the findings of the Royal Commission on Aboriginal Peoples had been largely ignored in 1996. From a different standpoint, it may be argued that the government's minimal participation was in fact an open statement for its potential voters, and therefore a clear assertion for those Canadian citizens who would not favor changing institutional relations with indigenous peoples. In any case, a contrast was evident between the TRC's 94 recommendations calling for immediate action and the government's restrained participation in the TRC recommendations release.

As a member of the audience on June 2, 2015 in Winnipeg during the announcement of the recommendations, the relevance of various symbolic actions was palpable. We watched the day begin in Ottawa and Winnipeg, with the sacred fire, left to burn until the end of the final ceremonial event. The emotional side of the celebration was as important as the rational one. Grief was passed on from person to person. The members of the TRC themselves had clearly experienced shock and distress while listening to the reports and collecting the terrible stories. This point was acknowledged in their individual statements[76]. The underlying message travelled around the room and across the country, linking past and present histories at all levels of the community, touching first and second-generation survivors, and expanding to include a community of participants. The histories implicitly referred to were personal, familiar, collective, and national stories of subjection and despair, remembrance and hope, recognition and action, celebration and joy. Photos published in newspapers showed survivors holding their children and

74 See Curry and Galloway (2015: 6): "'We are still awaiting the full report. The government will examine all of these and, obviously, read them before deciding what the appropriate next steps are', Mr Harper told the House of Commons".

75 "On June 2, 2015, the Government of Canada and the other parties to the Settlement Agreement received the TRC's executive summary, including its key findings and recommendations. The government will study these findings and recommendations carefully. We look forward to receiving the entire six volume final report to be able to fully understand the TRC's conclusions and respond in an informed manner."

76 Moran (2015: 11).

grandchildren[77]. Ceremonies are essential in indigenous cultures, and the TRC release was meant to be a collective national ceremony, too. Public spaces were occupied in order to infuse them with collective meaning. In Ottawa, "a symbolic show of reconciliation" took place on 1 June with a walk across the city[78].

On June 2, 2015 in Winnipeg, the official release was broadcast at the University of Winnipeg, where a crowd of about 700 people had gathered to witness the event. In attendance were first and second-generation survivors of residential schools, as well as descendants, politicians, institutional representatives, academics, and supporters like myself. Of the several local speakers, Grand Chief Derek Nepinak of the Assembly of Manitoba Chiefs delivered an outspoken and provocative speech, claiming that "[t]he atrocities of yesterday are happening right now", which was clearly challenging political institutions to take action and "make changes within this lifetime", in order to accomplish the TRC's work and the struggle of indigenous peoples in Canada[79]. Winnipeg celebrations continued with a march along the public urban spaces of Portage Avenue and Main Street, as far as the Thunderbird House, where a ceremonial pipe smoking ceremony and singing took place. Members of the public in attendance were concerned and emotional. People were respectful and appreciative towards the survivors, who had endured terrible struggle, and now took part in the celebration showing both their pride and sorrow. External observers, like myself, could perceive the distress, the inexplicability of the history being resuscitated, and the need for new stories to be articulated. However, the general climate was one of festivity combined with constructive anticipation for effective action to be devised on the political, social and cultural arenas.

The TRC's proposal for future action entails teaching and education. On the TRC website, teaching is presented as a permanent project, a lifestyle, a necessary tool for survival, exchange, individual and collective growth[80]. To this end, the protection and preservation of indigenous languages has been a major concern. The

77 See Curry and Galloway (2015: 6): "Former Northwest Territories premier Stephen Kakfwi, a residential school survivor, holds his granddaughter Sadeya Kakfwi-Scott while standing with the audience at the Truth and Reconciliation Commission on Tuesday in Ottawa"; see Rabson (2015: 3): "Charlotte Boubard holds her daughter, Toni, while singing at the Winnipeg's Thunderbird House after the Walk for Reconciliation on Tuesday"; see Welch (2015: 4): "Caroline Ouskun and her nephew Travis Spence [...] at the U of W".

78 See Rabson (2015: 3).

79 See Welch (2015: 4).

80 www.trc.ca.

oral recording and written transcription of native languages is being undertaken by experts in the field. In order to raise awareness, elders speak in public about their versions of history and cultural practices important to them and characteristic of a good life. This is part of the rewriting of history that the TRC has envisaged. Literacy for indigenous children is demanded, too, so that young people may have access to information and resources that would otherwise be inaccessible to them.

The website "Truth and Reconciliation Commission of Canada" contains statements, documents, and research materials gathered by the Commission, which will be stored in the National Centre for Truth and Reconciliation[81]. The website is rich and detailed, up-to-date, and user-friendly. Sharing information and attracting engagement from the community are visible objectives. The heading "Findings" leads one to the executive summary, "Honouring the Truth, Reconciling for the Future", the principles "What We Have Learned. Principles of truth and reconciliation", the document "The survivors speak", and "Truth and Reconciliation Commission of Canada: Calls to Action". Examples show how simple gestures may become meaningful. "It matters to me" brings to the fore expressions of reconciliation, and the "Personal ribbon campaign" proposes the exchange of symbolic gifts. Invitations for "Sharing the truth" and "What can I do?" give detailed instructions for statement gathering. And activities are suggested for groups of people and for children in order to mark a symbolic beginning for teaching a new history[82]. Emotionally and rationally, the website is engaging and thought provoking.

5. Coda: The Land We Are

In the inspiring climate around the release of the TRC findings summary, the timely book *The Land We Are*, published just after the release of the TRC summary, is worth mentioning because it warns against the myth of historical transformation. It points to the dangers of patronizing, top-down manipulation and romantic illusions of fulfilment, however appealing they might seem.

Postcolonial theory favors resistance, marginal points of view, and dissenting attitudes. They are essential theoretical and methodological tools that allow multiple voices and perspectives to engage actively with one another in context.

---

81 <http://umanitoba.ca/nctr/>.

82 Information about the Truth and Reconciliation Commission can be found on the websites of the federal government of Canada at <http://www.pm.gc.ca/>; <https://www.gg.ca/>; <https://www.aadnc-aandc.gc.ca/>; and on the website of the Canadian Broadcasting Corporation at <http://www.cbc.ca/news/canada/a-history-of-residential-schools-in-canada-1.702280>.

From this standpoint, the book offers a challenging reading of the notion of reconciliation in contemporary Canada, while putting forward ambitious solutions. Visual artists and performers in *The Land We Are* refute the idea of reconciliation when it is aimed at moving beyond and setting aside a problematic chapter of history. Instead, they advocate reparations and restitution as integral steps towards reconciliation[83]. Otherwise, they argue, reconciliation could simply instantiate acceptance of the status quo. Their criticism stems from various artistic fields. With visual installations, performances, and creation of artworks, artists criticize the public use of indigenous artistic traditions for gentrified urban developments, reclaiming public areas as meeting spaces for indigenous peoples. Their works also deconstruct institutional discourses, and propose collaborative practices that use the transformative and transgressive power of art and performance. The book articulates an open statement mainly addressed to non-indigenous Canadians who have the real and unavoidable responsibility to abandon the normalizing colonial narrative of history from which they benefited, and to change, both practically and radically, their relationships with indigenous peoples by giving up privilege and power. For artists in *The Land We Are*, reconciliation in Canada is just a beginning. It can only progress if non-indigenous people learn to live with and collaborate with indigenous residents of their country.

In conclusion, the debate around the release of the summary of the findings of the Truth and Reconciliation Commission on Indian Residential Schools in Canada has not been examined as a direct example of neo-managerialism. However, it raises important questions concerning the type of citizens to whom public communication is addressed, as well as those citizens' place in society and in the national community. If an inclusive society appears to be based on transparent communication by the public central administration – which should set the basis for a new deal between the state and the citizens – then the example offered here is a key case for understanding how problematic and blurred the category of citizenship (and the related active participation of citizens in the civil society) may be in countries with a colonial history as a settler colony in inhabited lands. One should also be alert to the dangers that Larry Terry points out when he focuses on challenges to democracy if the state takes on the characteristics of a firm or of a futuristic e-democracy in countries where education, literacy, and basic social welfare for indigenous people are issues to be urgently addressed. It seems that a contradiction is inevitable when public administrations model themselves on corporations while also claiming to articulate long-term idealistic objectives of

83 See Hill and McCall (2015). See also, Dudley (2015: 1).

inclusive democracy. It must be further explored and verified whether addressing citizens as stakeholders through effective communication will allow the different voices of civil society to be represented. In addition, a vigilant attitude should be maintained so as to never to lose sight of the ways in which race and class act as formidable driving forces of marginalization, particularly when hegemonic neoliberal economic regimes control political, social, and cultural interactions. Surreptitious and ambiguous dynamics of social ostracism and cultural isolation must be assiduously pursued and exposed by speaking the truth against homogenizing neo-imperialist discourses and practices. This is beneficial advice given by postcolonial theorist Edward Said that helps bring dissenting and minority views into the public discourse. To this end, the Summary of the Truth and Reconciliation Commission of Canada offers precise indications and suggestions about the nature of social obligations, citizenship, and minority rights that extend beyond the contextual frame of the Commission's field of investigation into the organized abuses of indigenous peoples. Reading the extended final report, delivered on December 15, 2015[84], brings to the fore relevant lines of analysis and reflection, as Jocelyn Thrope suggests in the epigraph to this paper that Canadians "By listening and learning [...] can work together to make the changes our society desperately needs". It is a considerate recommendation that may also effectively inspire more general, perceptive and receptive attitudes of cultural awareness beyond the contingency of the case examined in this essay. Expectations and hopes about the real possibility for a cultural, social, and political change in Canada have risen since Justin Pierre James Trudeau won the federal elections held on October 19, 2015, becoming the Prime Minister of a majority liberal government that has been leading the country since November 4, 2015. His government set the indigenous question at the core of its political agenda.

References

Andersen, C. (2014). *Métis. Race, Recognition, and the Struggle for Indigenous Peoplehood.* Vancouver: UBC Press.

Anholt, S. (2009). "The media and national image". *Place Branding and Public Diplomacy* 5(3): 169–179.

84 <http://www.trc.ca/websites/trcinstitution/index.php?p=3>, <http://www.trc.ca/websites/trcinstitution/index.php?p=890>, <http://www.trc.ca/websites/trcinstitution/index.php?p=903> (last accessed March 5, 2016) and <http://nctr.ca/map.php> (last accessed March 8, 2016).

Aronczyk, M. (2013). *Branding the Nation. The Global Business of National Identity.* Oxford: Oxford University Press.

Bailey, S. (2015). "Premiers commit to commission recommendations after meeting with native leaders". <http://www.winnipegfreepress.com/canada/native-leaders-to-discuss--violence-children-in-care-at-meeting-with-premiers-315246151.html>.

Bandelj, N. and F. Wherry (eds) (2011). *The Cultural Wealth of Nations.* Stanford: Stanford University Press.

Banting, K., T. Courchene and F. Leslie Seidle (eds) (2007). *Belonging? Diversity, Recognition and Shared Citizenship in Canada.* Montreal: Institute for Research on Public Policy.

Banting, K. and W. Kymlicka (2010). "Canadian Multiculturalism: Global Anxieties and Local Debates". *British Journal of Canadian Studies* 23(1): 43–72.

Bryce, Henderson P. (1922). *The Story of a National Crime: Being an Appeal for Justice to Indians of Canada.* Ottawa: James Hope & Sons.

Campbell, D. (2015). "Metis, craft, civic mindedness: Essential attributes of democratic citizenship in communities". *Community Development,* 46(3): 198–211.

Carson, K. Blaze (2015). "History of violence". *The Globe and Mail:* 1–8.

Chiose, S. (29 May 2015). "Universities must address residential schools legacy, Justice Sinclair said", *The Globe and Mail*: <http://www.theglobeandmail.com/news/national/education/universities-must-address-residential-school-legacy-justice-sinclair-says/article24709958/>.

Chiose, S. (2015). "Education a way to reconciliation: justice". *The Globe and Mail*: 5.

Clifton, R. E. and H. Rubenstein (2015). "Debunking the half-truths and exaggerations in the Truth and Reconciliation report". *National Post,* <http://news.nationalpost.com/full-comment/clifton-rubenstein-debunking-the-half-truths-and-exaggerations-in-the-truth-and-reconciliation-report>.

Coates, K. (2015). "McLachlin said what many have long known". *The Globe and Mail*: 11.

Curry, B. (2015). "PM silent on recommendations at closing ceremony". *The Globe and Mail*: 1–9.

Curry, B. and G. Galloway. (2015). "A call to action". *The Globe and Mail*: 6–7.

D'Arcy, L., H. Bone, and the Treaty & Dakota Elders of Manitoba with contributions by the AMC Council of Elders (2014). Ka'esi Wahkotumahk Aski. Our Relations with the Land, Vol. 2, Winnipeg, Treaty Relations Commission of Manitoba & Assembly of Manitoba Chiefs Secretariat.

Davin, N. (1879). "Flood, Report on Industrial Schools for Indians and Half-Breeds, 1843–1901". <https://archive.org/details/cihm_03651>.

Day, R. J. F. (2000). *Multiculturalism and the History of Canadian Diversity*. Toronto: University of Toronto Press.

Fine, S. (2015). "McLachlin: A history of 'cultural genocide'". *The Globe and Mail* 2: 1–12.

Fine, S. (2015). "Chief Justice resets the native question". *The Globe and Mail*: 6.

Galloway, G. (2015). "Commission to chart map of rocky road to reconciliation". *The Globe and Mail*: 6–7.

Galloway, G. (2015). "The road map". *The Globe and Mail*: 1–8.

Galloway, G. and B. Curry (2015). "Cultural genocide' cited as goal of residential Schools". *The Globe and Mail*: 1–10.

Galloway, G. and B. Curry (2015). "Truth and Reconciliation's urgent call for political action". *The Globe and Mail*: 1, 8.

Garcea, J., A. Kirova, and L. Wong (2008). "Introduction: Multiculturalism Discourses in Canada". *Canadian Ethnic Studies Journal* 40(1): 1–10.

Geary, A. (2015). "Professors condemn colleagues. 'Retrograde' opinions on TRC denounced". *Winnipeg Free Press*, Editorial: 13.

The Globe and Mail, Editorial (June 3, 2015). "Genocide or not, we are the authors": 10.

Gould, N. (2015). "With academic freedom comes responsibility". *Winnipeg Free Press*: 9.

Government of Canada. *Report of the Special Committee of the House of Commons examining the Indian Act amendments of 1920 (Duncan Campbell Scott testimony on March 30, 1920)*, Library and Archives Canada, RG10, volume 6810, file 470–2–3, part 7.

Ibbitson, J. (2015). "Our attitudes need to change to make a difference". *The Globe and Mail*: 7.

Kymlicka, W. (2012). *Multiculturalism: Success, Failure, and the Future*. Washington, DC: Migration Policy Institute.

Kirkup, K. (2015). "Hard work' ahead to reconcile". *Winnipeg Free Press*: 4.

Kopecky, A. (2015). "Title Fight. The Tsilhqot'in Nation got its land back. Canada will never be the same". *The Walrus*: 30–39.

Kusch, L. (2015). "Harper's silence over TRC report confusing: Fontaine". <http://www.winnipegfreepress.com/local/Harpers-silence-over-TRC-report-confusing-Fontaine-307653241.html>.

Hill, G. and S. McCall (eds) (2015). *The Land We Are*. Winnipeg: ARP Books.

Ibbitson, J. (2015). "Our attitudes need to change to make a difference". *The Globe and Mail*: 7.

Lett, D. (2015). "A mountain waiting to be climbed". *Winnipeg Free Press*: 5.

Lett, D. (2015). "Are we up to the challenge of rebooting history?" *Winnipeg Free Press*: 4.

Lett, D. (2015). "Apology welcome, but doesn't excuse approach of controversial TRC column". *Winnipeg Free Press*. <http://www.winnipegfreepress.com/opinion/columnists/Apology-welcome-but-doesnt-excuse-condemned-editorial--311029771.html>.

Martin, N. (2015). "Several groups' signs on to language preservation agreement, but not the federal gov't". <http://www.winnipegfreepress.com/local/Mutlitude-of-groups-signs-on-to-preserve-foster-indigenous-languages-328513801.html>.

Miller Granville, B. (2003). *Invisible Indigenes. The Politics of Nonrecognition*. Lincoln: University of Nebraska Press.

Miller Granville, B. (2011). *Oral History on Trial. Recognizing Aboriginal Narratives in Courts*. Vancouver: University of British Columbia Press.

Mitchell, C. (2015). "Sixties Scoop apology just a start". *Winnipeg Free Press*: 9.

Moran, R. (2015). "What it really means to work together". *The Globe and Mail*: 11.

Muller, A. (2015). "Indian residential schools were unqualified genocide". *Winnipeg Free Press*: 9.

Pankratz, C. (2015). "Genocide that dares not speak its name". *Winnipeg Free Press*: 9.

Pasquin, Mali I. (2015). "Canada confronts its dark history of abuse in residential schools". <http://www.theguardian.com/world/2015/jun/06/canada-dark-of-history-residential-schools> (accessed 29 September 2015).

Perry, A. (2015). "This history is not over". *Winnipeg Free Press*: 9.

Pratt, D., H. Bone and the Treaty & Dakota Elders of Manitoba with contributions by the AMC Council of Elders (2014). Untuwe Pi Kin He. *Who We Are. Treaty Elders' Teachings*, Vol. 1. Winnipeg: Treaty Relations Commission of Manitoba & Assembly of Manitoba Chiefs Secretaria.

Puxley, C. (2015). "Sixties Scoop adoptees to get provincial apology". *Winnipeg Free Press*: 10.

Puxley, C. (2015). "Métis fear they are being side-lined". *Winnipeg Free Press*: 4.

Rabson, M. (2015). "Canadians united for change". *Winnipeg Free Press*: 3.

Rabson, M. (2015). "To remember and to change". *Winnipeg Free Press*: 3.

Rabson, M. (2015). "New poll shows many Canadians not paying attention to issue of residential schools". <http://www.winnipegfreepress.com/canada/

New-poll-shows-many-Canadians-not-paying-attention-to-issue-of-residential-schools-312964481.html>.

Rabson, M. (2015). "Cultural genocide' term gets backing". *Winnipeg Free Press*: 12.

Sampert, S. (2015). "Aboriginals may lose patience if report doesn't stir action". *Winnipeg Free Press*: 9.

Simpson, J. (2015). "Fixating on past is not productive". *The Globe and Mail*: 11.

Swan, M. (2015). "Battle never over for school survivors". *Winnipeg Free Press*: 3.

Terry, Larry D. (1998). "Administrative Leadership, Neo-Managerialism, and the Public Management Movement". *Public Administration Review* 58(3): 194–200.

Thorpe, J. (2015). "We can start to reconcile now". *Winnipeg Free Press*: 9.

Towtongie, C. and M. Coon Come (2015). "Canada is shirking its duty to its people". http://www.winnipegfreepress.com/opinion/analysis/canada-is-shirking-its-duty-to-its-people-327316191.html.

The Truth and Reconciliation Commission, Mandate, http://www.trc.ca/websites/trcinstitution/File/pdfs/SCHEDULE_N_EN.pdf.

The Truth and Reconciliation Commission of Canada, Honoring the Truth, Reconciling for the Future. Summary of the final report of the Truth and Reconciliation Commission of Canada, 2015, www.trc.ca.

The Truth and Reconciliation Commission of Canada, Truth and Reconciliation Commission of Canada: Calls to Action, 2015, www.trc.ca.

The Truth and Reconciliation Commission of Canada, What We Have Learned: Principle of Truth and Reconciliation, 2015, www.trc.ca.

The Truth and Reconciliation Commission of Canada, The Survivors Speak: A Report of the Truth and Reconciliation Commission of Canada, 2015, www.trc.ca.

'Truth and consequences'. *The Economist*, June 6, 2015: 28.

'Truth and Reconciliation's urgent call for political action'. *The Globe and Mail:* 1.

United Nations Convention on the Prevention and Punishment of the Crime of Genocide, 1948, <https://treaties.un.org/doc/Publication/UNTS/Volume%2078/volume-78-I-1021-English.pdf>.

United Nations Declaration on the Rights of Indigenous Peoples, 2007, <http://www.un.org/esa/socdev/unpfii/documents/DRIPS_en.pdf>.

Welch, M. A. (2015). "Battle never over for school survivors". *Winnipeg Free Press*: 3.

Welch, M. A. (2015). "Colonial patterns are still there". *Winnipeg Free Press*: 4.

Winnipeg Free Press Editorial (2015). "Will truth set us free?" *Winnipeg Free Press*: 9.

Winnipeg Free Press Editorial (June 1, 2015). "Still taking 'Indian' out of the child?" *Winnipeg Free Press*: 8.

Winnipeg Free Press Editorial (June 3, 2015). "The Search for healing and justice". *Winnipeg Free Press*: 7.

Winnipeg Free Press Editorial (June 17, 2015). "Silence on TRC 'confusing'" *Winnipeg Free Press*.

Winnipeg Free Press Editorial (2015). "Words are not enough. Reconciliation requires deliberate, thoughtful and sustained action". *Winnipeg Free Press*: 1.

Sites

Idle No More, <http://www.idlenomore.ca>

Indigenous and Native Affairs Canada, <https://www.aadnc-aandc.gc.ca/eng>

Manitoba Museum, <www.manitobamuseum.ca>

Treaty Relations Commission of Manitoba, <http://www.trcm.ca/>

www.ingramcontent.com/pod-product-compliance
Ingram Content Group UK Ltd.
Pitfield, Milton Keynes, MK11 3LW, UK
UKHW041828100926
8556IPUK00002B/71

9 783631 666838